PROBABILISTIC FRACTURE MECHANICS AND FATIGUE METHODS: APPLICATIONS FOR STRUCTURAL DESIGN AND MAINTENANCE

A symposium
sponsored by ASTM
Committees E-9 on Fatigue
and E-24 on Fracture Testing
St. Louis, Mo., 19 Oct. 1981

ASTM SPECIAL TECHNICAL PUBLICATION 798
J. M. Bloom, Babcock and Wilcox Co.
J. C. Ekvall, Lockheed-California Co.
editors

ASTM Publication Code Number (PCN)
04-798000-30

 1916 Race Street, Philadelphia, Pa. 19103

Copyright © by AMERICAN SOCIETY FOR TESTING AND MATERIALS 1983
Library of Congress Catalog Card Number: 82-83518

NOTE

The Society is not responsible, as a body,
for the statements and opinions
advanced in this publication.

Printed in Baltimore, Md. (b)
June 1983

Foreword

This publication, *Probabilistic Fracture Mechanics and Fatigue Methods: Applications for Structural Design and Maintenance,* contains papers presented at the symposium on Probabilistic Methods for Design and Maintenance of Structures which was held in St. Louis, Missouri, 19 Oct. 1981. The symposium was sponsored by ASTM Committees E-9 on Fatigue and E-24 on Fracture Testing. J. M. Bloom, Babcock and Wilcox, Co., and J. C. Ekvall, Lockheed-California Co., presided as symposium chairmen and editors of this publication.

Related ASTM Publications

Residual Stress Effects in Fatigue, STP 776 (1982), 04-776000-30

Low Cycle Fatigue and Life Prediction, STP 770 (1982), 04-770000-30

Design of Fatigue and Fracture Resistant Structures, STP 761 (1982), 04-761000-30

Methods and Models for Predicting Fatigue Crack Growth Under Random Loading, STP 748 (1981), 04-748000-30

Fracture Mechanics (13th Conference), STP 743 (1981), 04-743000-30

Fractography and Materials Science, STP 733 (1981), 04-733000-30

A Note of Appreciation to Reviewers

The quality of the papers that appear in this publication reflects not only the obvious efforts of the authors but also the unheralded, though essential, work of the reviewers. On behalf of ASTM we acknowledge with appreciation their dedication to high professional standards and their sacrifice of time and effort.

ASTM Committee on Publications

ASTM Editorial Staff

Janet R. Schroeder
Kathleen A. Greene
Rosemary Horstman
Helen M. Hoersch
Helen P. Mahy
Allan S. Kleinberg
Virginia M. Barishek

Contents

Introduction   1

PROBABILISTIC FRACTURE MECHANICS

Probabilistic Evaluation of Conservatisms Used in Section III, Appendix G, of the ASME Code—G. M. JOURIS   7
   Discussion   17

Applications of a Probabilistic Fracture Mechanics Model to the Influence of In-Service Inspection on Structural Reliability—D. O. HARRIS AND E. Y. LIM   19

Statistical Scatter in Fracture Toughness and Fatigue Crack Growth Rate Data—G. O. JOHNSTON   42

Probabilistic Defect Size Analysis Using Fatigue and Cyclic Crack Growth Rate Data—G. G. TRANTINA AND C. A. JOHNSON   67

Statistical Methods for Estimating Crack Detection Probabilities—A. P. BERENS AND P. W. HOVEY   79

STATISTICAL ASPECTS OF FATIGUE

Characterization of the Variability in Fatigue Crack Propagation Data—D. F. OSTERGAARD AND B. M. HILLBERRY   97

Exploratory Study of Crack-Growth-Based Inspection Rationale—E. K. WALKER   116

Cumulative Damage: Reliability and Maintainability—F. KOZIN AND J. L. BOGDANOFF   131

Method for Determining Probability of Structural Failure from Aircraft Counting Accelerometer Tracking Data—C. E. LARSON AND W. R. SHAWVER   147
   Discussion   160

Analysis of Structural Failure Probability Under Spectrum Loading Conditions—C. E. BRONN   161

***S-N* Fatigue Reliability Analysis of Highway Bridges**—
PEDRO ALBRECHT   184

Summary   205

Index   213

Introduction

This Special Technical Publication is a collection of papers approved for publication in connection with the Symposium on Probabilistic Methods for Design and Maintenance of Structures jointly sponsored by ASTM Committee E-9 on Fatigue and Committee E-24 on Fracture Testing in St. Louis, Mo., on 19 Oct. 1981. The purpose of that symposium was to create a forum for the presentation and discussion of probabilistic methodology used to determine structural safety and reliability. Structural reliability is interconnected with safety of structure before service through design and after service through maintenance.

Structural safety and reliability are a function of many variables, such as:

1. Fatigue, crack growth, and fracture properties of materials and structures.
2. Variations of environment and service loadings.
3. Type and frequency of production and in-service inspections.

Therefore, probabilistic methods are needed to characterize these variables and the interrelationship of these variables for the design and maintenance of structures.

With the development of increasing complex structural systems, the control of the quality of the components and system operation becomes exceedingly more difficult. The overall high reliability and availability desired becomes difficult to achieve or guarantee. During World War II, the need for reliability analyses was first recognized in the electronics industry. The idea of reliability analyses was developed for the resolution of accidents that resulted from complexity, sophistication, and newness of the electronics industry.

With regard to structure, satisfactory safety and reliability have been achieved over a period of time by changing design criteria based on service experience. For example, in the aircraft industry, safe-life design criteria were implemented to improve safety and reliability due to failures caused by fatigue cracking. This criteria proved unsatisfactory, and the fail-safe design approach was required for the design of commercial aircraft in 1956. The safety record was considerably improved for structure designed to this criteria, but some failures still occurred due to undetected fatigue cracking. Now damage tolerance design criteria are required that must provide damage growth rates that are compatible with the inspection program. Also in some instances, the safety of a damage tolerance design may be assessed by probabilistic evaluation employing methods such as risk analysis.

Deterministic methods have been generally used to assess structural reliability or safety. In the deterministic approach, safety can be measured in terms of a safety factor or margin of safety. The safety factor for structures is usually determined by the ratio of a strength parameter divided by a stress parameter. These parameters have been typically based on average values. Other definitions involve lower bounding the strength parameter and upper bounding the stress parameter. Both these definitions, however, ignore the fact that these parameters are statistically distributed. In many industries, for example, the nuclear power industry, these deterministic values used for strength and stress are chosen as worst case quantities that, when considered in total, give unrealistically conservative assessments of safety. The chances that these worst case values can occur simultaneously is extremely low. What is needed is a probabilistic approach to quantify the degree of inherent safety of the structures. The probabilistic measure of safety is reliability that is a statistical measure where for structures—governing strength exceeds governing stress causing failure.

To use probabilistic or risk analysis methods requires sufficient data to statistically characterize the variables that affect structural reliability and safety. Also, methodology needs to be developed for relating not only strength and stress parameters, but consideration of the damage condition of the structures, the rates of damage growth, and inspections performed to detect damage before it reaches catastrophic proportions. The papers presented in this book address some aspect of the probabilistic methodology used to determine structural safety and reliability.

The symposium, as well as this publication, were divided into two categories of papers; although it is recognized in some cases there is overlap. The categories are fracture and fatigue, or stated another way: Probabilistic Fracture Mechanics (PFM) and Statistical Aspects of Fatigue.

In 1979, ASTM Committee E-24 decided that there was enough interest to establish a task group (E24.06.03) under E24.06, the application of fracture mechanics group, to develop and implement PFM. The scope of this new group included consideration of all variables needed in a fracture mechanics probabilistic analysis; namely, the probabilistic distribution of defects in a structure (initial quality), the probabilistic distribution of material properties (fracture toughness and fatigue crack growth rates), and the probabilistic distribution of loadings (static and fatigue) applied to the structures. These distributions are then used to determine the reliability of the structure based on a brittle or ductile fracture mechanism.

Committee E-9 Task Groups on Statistical Planning of Fatigue Tests and on Nomenclature, founded in 1951 and 1959, respectively, were combined in 1960 to form Subcommittee E09.06 on Statistical Aspects of Fatigue. The primary goal of this subcommittee has been to bring probabilistic and statistical methods of experimental design, data gathering, and analysis within the grasp of the practicing engineer. To achieve these goals the subcommittee sponsors workshops, symposia, and tutorial lectures for the dissemination of new and current information on the statistical aspects of fatigue.

Two recent accomplishments of Task Group E24.06.03 and Subcommittee E09.06 include the following:

1. Special Program on Probabilistic Fracture Mechanics, Pittsburgh, Pa., 31 Oct. 1979. This program consisted of five speakers who discussed application of PFM to a wide range of structures, for example, aircraft, bridges, automobiles, piping, and pressure vessels.
2. A Symposium on Statistical Analysis of Fatigue Data held on 30-31 Oct. 1979 in Pittsburgh, Pa. Six papers presented at this symposium and ASTM Standard E 739-80 are published in Special Technical Publication 744, August 1981.

The papers presented in this book are representative of the state-of-the-art application of probabilistic aspects of fracture mechanics and fatigue. It is hoped that these papers will expose the practicing design engineer, as well as those engineers who must evaluate the safety of structures, to the probabilistic approach.

This Special Technical Publication is sponsored by ASTM Committee E-24 on Fracture Testing and Committee E-9 on Fatigue. We acknowledge the contributions made by the authors, reviewers, and ASTM staff.

J. M. Bloom

Babcock & Wilcox, Research and Development Division Alliance, Ohio; symposium cochairman and editor.

J. C. Ekvall

Lockheed-California Company, Burbank, Calif.; symposium cochairman and editor.

Probabilistic Fracture Mechanics

G. M. Jouris[1]

Probabilistic Evaluation of Conservatisms Used in Section III, Appendix G, of the ASME Code

REFERENCE: Jouris, G. M., **"Probabilistic Evaluation of Conservatisms Used in Section III, Appendix G, of the ASME Code,"** *Probabilistic Fracture Mechanics and Fatigue Methods: Applications for Structural Design and Maintenance, ASTM STP 798,* J. M. Bloom and J. C. Ekvall, Eds., American Society for Testing and Materials, 1983, pp. 7-18.

ABSTRACT: In this paper the results of a probabilistic analysis for a pressure vessel are described in which the effects of various conservatisms used in Section III, Appendix G, of the ASME Boiler and Pressure Vessel Code are assessed. The methodology used is based on the introduction of probabilistic concepts into the deterministic calculations of Section III, Appendix G, so that the appropriate variability is reflected. By comparing the estimated probabilities of failure based on the various conservatisims, the approximate margin of safety in the design of a pressure vessel has been evaluated. From this analysis, it is concluded that this safety margin is considerable. Depending upon the conditions considered to be realistic in practice, the margin could be upwards of 10 or more orders of magnitude.

KEY WORDS: probabilistic fracture mechanics, conservatisms, test simulation, safety factor, stress intensity, fracture toughness, pressure vessel, flaws, failure, fatigue (materials), fracture mechanics

In this paper some results of a probabilistic analysis are described in which the effects of various conservatisms used in Section III, Appendix G, of the ASME Boiler and Pressure Vessel Code are assessed.

In the deterministic analysis based on Section III, Appendix G, the purpose is to show that the stress intensity factors for various parts of the pressure vessel at various operating and testing conditions will always be within the limit of the fracture toughness of the vessel material at various temperatures in order to ensure the protection against nonductile failure of the pressure vessel [1].[2] A margin of safety is introduced in the design of the ves-

[1] Senior statistician, Westinghouse R&D Center, Pittsburgh, Pa. 15235.
[2] The italic numbers in brackets refer to the list of references appended to this paper.

sel by virtue of the methods prescribed in Section III, Appendix G. The resulting overdesign of the pressure vessel is due in part to the following code assumptions:

1. Applied loads are used that are significantly larger than would be experienced in practice.
2. Fracture toughness values are used that are lower bound values.
3. A flaw is assumed that is larger than any likely to be present.

Using probabilistic fracture mechanics, the effects of these pessimistic assumptions on the probability of brittle failure of the pressure vessel have been assessed. The methodology used in the probabilistic analysis for assessing the effects of conservatisms, is based on the introduction of probabilistic concepts into the deterministic calculations of Section III, Appendix G, so that appropriate variability is reflected. In carrying out this design evaluation analysis, the various inputs to the analysis have been categorized either as constant values or random variables. For the random variables, the form of the variability has been reflected by probability density functions. Further, the various severity factors are identified and input as constants.

It must be emphasized that we are using the resulting probability estimates in a comparative rather than an absolute way. We do not believe that the true probability of failure is exactly that estimated but simply that it is an estimate that we will use to compare with other estimates. Our hesitation to use the estimates in an absolute way stems mainly from the facts that these estimates are highly dependent on the tail behavior of the assumed distributions, we have only limited data on which to base these, and the initial flaw distribution is not reflected here.

Using this methodology, the estimated probability of failure under the code assumptions is 10^{-11} for an example pressure vessel. If, however, we use more realistic applied loads, the estimated probability of failure is 10^{-20}. If we use a more realistic flaw size, the estimated probability of failure is 10^{-23}. One can see the degree of conservatism by comparing these probability estimates.

The analysis has been carried out for a variety of transients occurring in different regions of the pressure vessel. In all cases, the analysis indicates the design based on the code is very conservative.

Probabilistic Methodology

The methodology used in developing the computer program for assessing the effects of severity factors and conservatisms is based on the introduction of probabilistic concepts into the deterministic ASME Code, Section III, Appendix G, so that appropriate variability is reflected. In carrying out this analysis, the various inputs have been categorized either as constant values or random variables. For the random variables, the form of the variability will be reflected by a probability density function. Further, the various severity factors are identified and input as constants. Based on the values of the severity factors currently accepted, we have used the probabilistic code to get an estimate of the probability of failure for the component of interest. To assess the impact of a severity factor, we redo the analysis with the severity factor of interest assigned a more realistic value. By comparing the resulting estimate of probability of failure with the estimate obtained for the currently accepted value of the severity factor, we have a measure of the "margin of safety" attributable to this factor.

An appendix G analysis is an umbrella type of fracture mechanics analysis based on the principle of linear elastic fracture mechanics (LEFM). The purpose of this analysis is to show that the stress intensity factors for various parts of the pressure vessel at various operating and testing conditions will be always within the limit of the fracture toughness of the vessel material at various temperatures in order to ensure the protection against nonductile failure of the pressure vessel. The probability of failure will be expressed as the probability that the stress intensity, K_I, exceeds the critical stress intensity, K_{IR}, of the material. K_I will be a random variable by virtue of the fact that it is obtained as a function of transient temperature and pressure that are assumed to be random variables. Both are taken to have normal distributions. K_{IR} is also a random variable because it is a function of the reference temperature shift (normal) and inherent variability (gamma) about the mean curve. See Appendix B for more details on the distributions of these random variables.

The development of the computer code to estimate the probability of failure for our comparative analysis was complicated by two competing forces. On the one hand, it was necessary to maintain a sufficient level of detail in the computer code to allow the various severity factors and conservatisms to be identified, but, on the other hand, the detail could not be so great (such as including a finite element analysis) that the computer time required was prohibitive. The computer time problem is critical because Monte Carlo simulation is used to obtain the probability estimates. This technique, virtually the only way to estimate a probability in a problem as complex as this, requires that the deterministic calculations be repeated about 5000 times for random settings of each of the input variables; a time-consuming process itself. In order to efficiently carry out the Monte Carlo simulation, importance sampling has been used (see Appendix I). This is a method for biasing the sampling by means of an importance distribution and then correcting for this bias in the probability calculation. It should be noted that in using importance sampling the results are affected by the choice of the importance density. The choice is more art than science.

It should be noted that in using importance sampling the results can be affected by the choice of the importance density, but since such change is unlikely to affect the comparisons made, it was not studied.

Results of Probability Estimation

This section contains the results of using the computer program to assess the effects of conservatisms and severity factors used in Section III, Appendix G, of the ASME Boiler and Pressure Vessel Code. Before discussing the probability estimates and making comparisons, we must specify the conditions under which the calculations are being made. Except where noted, we will use the following conditions.

1. Assume 32 effective full power years of life.
2. The presence of a 0.25 thickness surface flaw oriented in the longitudinal direction.
3. The use of Westinghouse trend curve of embrittlement shift [3].
4. The use of the standard lower bound K_{IR} curve for fracture toughness [1].
5. The multiplier γ will be set at 2.0 in the relationship

$$K_I = \gamma K_{Ip} + K_{It}$$

where K_I is the stress intensity, K_{Ip} is the primary (pressure) stress intensity factor, K_{It} is the secondary (thermal) stress intensity factor, and γ is a severity multiplier for K_{Ip}.

Beltline Region Analysis

We begin by estimating the probability of brittle failure in the beltline region (lower shell) of the pressure vessel for various transients. As can be seen in Table 1, the estimated probability of failure is fairly constant except for the heatup, cooldown, and turbine roll conditions. The higher prob-

TABLE 1—*Estimated probability of brittle failure for various transients, beltline region.*

Transient	Estimated Failure Probability
Heat up	8×10^{-7}
Cool down	6×10^{-7}
Loading	7×10^{-11}
Unloading	8×10^{-11}
Stepload increase	2×10^{-11}
Stepload decrease	7×10^{-11}
Large stepload decrease	1×10^{-11}
Loss of load	9×10^{-11}
Loss of power	9×10^{-11}
Loss of flow	2×10^{-10}
Reactor trip	4×10^{-10}
Turbine roll	1×10^{-8}
Steady state	4×10^{-10}

abilities in these cases are most readily explained by the lower temperatures present.

Nonbeltline Region Analysis

The failure probability was also estimated for other regions of the pressure vessel. Since the probabilities were so nearly the same for each of the transients in the beltline region, only a representative transient, stepload increase, was used for the outlet nozzle, lower head, and upper (closure) head regions. These probabilities are given in Table 2, and we see only fairly small differences from region to region.

Comparison of Severity Factors

We want to use the computer program to assess the impact of changes in the assumptions used in Section III, Appendix G, of the ASME Code. All the comparisons are for the beltline region with a stepload increase transient after 32 effective full power years of life. The other four conditions detailed at the beginning of this section will be varied. These options are as follows.

1. Multiplier γ in $K_I = \gamma K_{Ip} + K_{It}$: γ will vary from 1.2 to 2.0.
2. Flaw depth: This will vary from 0.15 thickness to 0.25 thickness.
3. Embrittlement shift: This will be based on the Westinghouse trend curve or the trend curve given in Regulatory Guide 1.99 [2].
4. Fracture toughness: This will be based on the standard lower bound K_{IR} curve or a mean curve with appropriate variability introduced [4].

In Fig. 1, one can see the effect on estimated failure probability of reducing the multiplier, γ (this has the effect of lowering the applied stress), and the hypothesized flaw depth to more realistic values. The vessel is designed for $\gamma = 2.0$ and assumes the presence of a 0.25 thickness flaw. The probability that a vessel so designed will fail under more realistic but still conservative conditions ($\gamma = 1.4$, 0.2 thickness flaw) is estimated to be more than 20 orders of magnitude less than was calculated for the design case ($\gamma = 2.0$, 0.25 thickness flaw). This is a measure of the level of conservatism intro-

TABLE 2—*Estimated probability of brittle failure for stepload increase transient.*

Region of Pressure Vessel	Estimated Probability of Brittle Failure
Beltline	2×10^{-11}
Outlet nozzle	9×10^{-12}
Lower head	1×10^{-11}
Upper head	5×10^{-12}

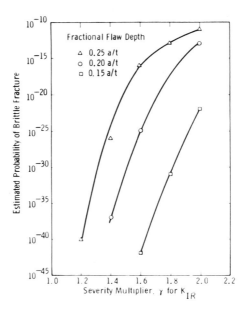

FIG. 1—*Change in estimated probability of failure as a function of γ and fractional flaw depth.*

duced by the ASME Code through this severity multiplier, γ, and the flaw depth. The results are based on the use of Westinghouse trend curve, mean fracture toughness curve, and stepload increase transient in the beltline region.

In Fig. 2, the use of the two types of irradiation embrittlement trend curves are compared. The W signifies the Westinghouse developed trend curve is used, while the 1.99 indicates the upper bound curve specified in Regulatory Guide 1.99 is used. We can see that use of the trend curve in Regulatory Guide 1.99 leads to a more conservative design than the use of the Westinghouse trend curve. The results are based on the use of mean fracture toughness curve, stepload increase transient in beltline region, and 0.25 thickness flaw. However, if $\gamma = 2.0$, the estimated probability of failure is about the same for the two curves. If instead we redo the calculations using the lower bound K_{IR} curve, the results are as shown in Fig. 3. Again, the results are based on a stepload increase transient in the beltline region containing a 0.25 thickness flaw. Again, we see the use of the curve in Regulatory Guide 1.99 leads to the more conservative design. However, the curves are quite close to each other, and it would appear that the relative difference is fairly small when γ is greater than 1.6 or so.

In order to compare how changing the fracture toughness curve affects the failure probability, we will reuse the results in Figs. 2 and 3 in a different

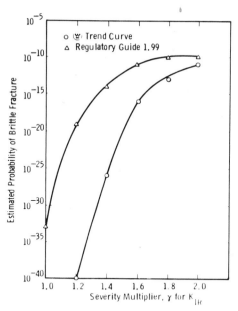

FIG. 2—*Change in estimated probability of failure as a function of γ and the trend curve used (mean fracture toughness curve used).*

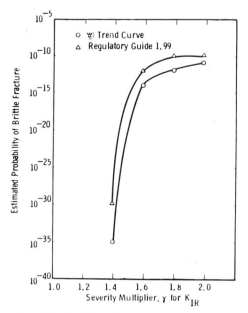

FIG. 3—*Change in estimated probability of failure as a function of γ and the trend curve used (lower bound K_{IR} curve used).*

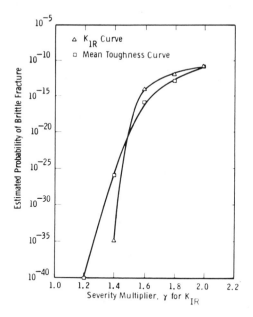

FIG. 4—*Change in estimated probability of failure as a function of γ and the fracture toughness curve used (Westinghouse trend curve used).*

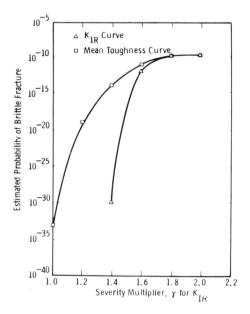

FIG. 5—*Change in estimated probability of failure as a function of γ and the fracture toughness curve used (Regulatory Guide 1.99 curve used).*

way. In Fig. 4, the comparison is made between the use of the standard lower bound K_{IR} curve and a mean fracture toughnesss curve with variability introduced. In this case, where the W trend curve has been used we can see that the more conservative curve depends on the γ chosen. However, in Fig. 5 where the two fracture toughness curves are again compared, but this time using the embrittlement shift curve from Regulatory Guide 1.99, the mean toughness curve is uniformly more conservative.

A wide variety of comparisons can be made, but those presented were judged to be of the greatest current interest. As the emphasis changes, other comparisons can be incorporated into the computer routine.

References

[1] ASME Boiler and Pressure Vessel Code, Section III, Appendix G, "Protection Against Nonductile Failure," American Society of Mechanical Engineers, New York, 1974, p. 487.
[2] United States Nuclear Regulatory Commission, Regulatory Guide 1.99, "Effects of Residual Elements on Predicted Radiation Damage to Reactor Vessel Materials," Revision 1, April 1977, pp. 1.99-3.
[3] Buchalet, C. B. and Bamford, W. H., "Method For Fracture Mechanics Analysis of Nuclear Reactor Vessels under Severe Thermal Transients," Paper 75-WA/PVP-3, Pressure Vessels and Piping Division, American Society of Mechanical Engineers, Winter Meetings, Dec. 1975.
[4] Jouris, G. M. and Shaffer, D. H., "Probabilistic Brittle Fracture Analysis For Major Thermal Transients in Pressure Vessels." *Proceedings*, 3rd International Conference on Structural Reliability in Reactor Technology, Paper G3/3 London, Sept. 1975.
[5] Hammersly, J. W. and Handscomb, D. C., *Monte Carlo Methods*, Methuen & Co., London, 1964.

APPENDIX I

Importance Sampling

Consider a random variable X with continuous density $f(x)$. The probability that X will lie below Z is

$$Pr(X \leq Z) = \int_{-\infty}^{Z} f(x)dx.$$

Let $I_Z(X)$ be defined by

$$I_Z(X) = \begin{cases} 1 & X \leq Z \\ 0 & X > Z \end{cases}$$

Then

$$Pr(X \leq Z) = \int_{-\infty}^{\infty} I_Z(x)f(x)dx = E(I_Z)$$

where the expectation is taken with respect to the random variable X.

Let $g(x)$ be a positive function subject to

$$\int_{-\infty}^{\infty} g(x)dx = 1$$

so that $g(x)$ can be interpreted as a probability density function. Now we write

$$\int_{-\infty}^{\infty} I_Z(x)f(x)dx = \int_{-\infty}^{\infty} I_Z(x)\frac{f(x)}{g(x)} g(x)dx = E\left(I_Z(x)\frac{f(x)}{g(x)}\right)$$

where the expectation is now taken with respect to a random variable whose density is $g(x)$. This new density, $g(x)$, is called the importance density as contrasted with $f(x)$, known as the parent density.

The straightforward and importance methods of evaluating $Pr(X \leq Z)$ are contrasted below.

Straightforward Monte Carlo	Importance Sampling
1. Draw a number from a population whose density is $f(x)$; call it x_1	1. Draw a number from a population whose density is $g(x)$; call it x_1
2. Evaluate $h_1 = I_Z(x_1)$	2. Evaluate $h_1 = I_Z(x_1)\dfrac{f(x_1)}{g(x_1)}$
3. Repeat Steps 1 and 2 for a succession of x_i	3. Repeat Steps 1 and 2 for a succession of x_i
4. Calculate $\hat{p} = \dfrac{1}{m}\sum_{i=1}^{m} h_i$	4. Calculate $\hat{p} = \dfrac{1}{m}\sum_{i=1}^{m} h_i$

The number of $\hat{p}$ obtained by following either procedure is an estimate of $Pr(X \leq Z)$. Using the straightforward method, most of the h values will be zero if the probability being estimated is quite small, while the h values from the importance sampling method will have a variety of small values. It can be shown that the variance of $\hat{p}$ from importance sampling can be made much less than the variance of $\hat{p}$ from the straightforward method, when the same m is used for each. In fact, a procedure, known as Spanier's algorithm, has been developed to determine an optimum importance density in the sense of minimum variance; the procedure is iterative.

The particular technique used to obtain the results quoted in the last two sections of this report is simply a higher dimensional version of that just shown. Our problem involved two input random variables. Let

$$X = \phi(W, Y)$$

be a relation that defines the random variable X in terms of two random variables W and Y, which respectively have densities $f_W(w)$ and $f_Y(y)$. We will denote by $f(x)$ the density of X, although initially that is not given. The probability that X does not exceed Z is given by

$$Pf(X \leq Z) = \int_{-\infty}^{Z} f(x)dx = \int_{-\infty}^{\infty} I_Z(x)f(x)dx = E(I_Z(x))$$

where I_Z is defined as before and the expectation is taken with respect to X. We may replace the single integration in x with an integration over the sample space as

$$\int_{-\infty}^{\infty} I_Z(x)f(x)dx = \int_{-\infty}^{\infty}\int_{-\infty}^{\infty} I_Z(x)f_W(w)f_Y(y)dwdy$$

again giving the expected value of $I_Z(x)$. Now introduce two importance densities, $g_W(w)$ and $g_Y(y)$, and so get

$$Pr(X \le Z) = \int \int I_Z(x) \frac{f_W(w) \cdot f_Y(y)}{g_W(w) \cdot g_Y(y)} g_W(w) \cdot g_Y(y) dwdy$$

$$= E\left[I_Z(x) \frac{f_W(w) \cdot f_Y(y)}{g_W(w) \cdot g_Y(y)}\right]$$

where now the expectation is taken relative to the importance densities.

APPENDIX II

Probability Distributions

The quantities listed here are random variables in the analysis with the distributions as specified.

Transient Temperature—This random variable will have a normal distribution with mean μ_T as obtained from the deterministic analysis; the variance σ_T^2 has been estimated as $-14°C$ (6°F).

Transient Pressure—This random variable will have a normal distribution with mean μ_p as obtained from the deterministic analysis; the variance σ_p^2 has been estimated as 0.4 ksi.

Reference Temperature Shift—This random variable is taken to have a normal distribution with mean μ_R as calculated using the deterministic model and variance $\sigma_R^2 = 400$ based on some surveillance and test reactor data.

Inherent K_{IR} Variability—From earlier work [4] the inherent K_{IR} variability will have a gamma distribution with parameters $\alpha = 50$, $\lambda = 0.55$, and $\gamma = 11$.

DISCUSSION

H. O. Fuchs[1] (*written discussion*)—Can you briefly describe the techniques by which very low probabilities ($\times 10^{-20}$) are derived from a much smaller number ($\times 10^4$) of simulation sums and give references to the literature? In particular, how sensitive are the results to assumptions about the (unknown) tails of the distribution curves for K, t, etc.?

---

[1] Mechanical Engineering Dept., Stanford University, Stanford, Calif. 94305.

G. M. Jouris (author's closure)—The technique of importance sampling is used to estimate very small probabilities by using a smaller number of trials than would be required by the standard Monte Carlo method. The mathematical justification for this technique is given in Appendix I of this paper. Additional references are provided below.

The sensitivity of the results as a function of the tail of the distributions has not been studied in this paper, but it is expected that the sensitivity is dependent on the shape of the tail.

Nagel, P. M., "Importance Sampling in Systems Simulation," Fifth Annual Conference on Reliability and Maintainability, 18-20 July 1966, New York, N.Y.

Pollyak, Yu G., "Estimation of Small Probabilities in Statistical Simulation of Systems," *Engineering Cybernetics*, 1971, pp. 342-349.

Schroder, R. J., "Fault Trees for Reliability Analysis," *Proceedings*, 1970 Annual Symposium on Reliability, 3-5 Feb. 1970, Los Angeles, pp. 198-205.

Spanier, J., "An Analytic Approach to Variance Reduction," *SIAM Journal of Applied Mathematics*, Vol. 18, 1970, pp. 172-190.

Spanier, J., "A New Multi-Stage Procedure for Systematic Variance Reduction in Monte Carlo," *Proceedings*, Conference on Reactor Mathematics and Applications, Idaho, CONF-710302, Vol. II, U.S. Atomic Energy Commission, 1971, pp. 760-770.

Walsh, J. E., "Questionable Usefulness of Variance for Measuring Estimate Accuracy in Monte Carlo Importance Sampling Problems," *Symposium on Monte Carlo Method*, H. A. Meyer, Ed., Wiley, 1956, pp. 141-146.

D. O. Harris[1] and E. Y. Lim[1]

Applications of a Probabilistic Fracture Mechanics Model to the Influence of In-Service Inspection on Structural Reliability

REFERENCE: Harris, D. O. and Lim, E. Y., **"Applications of a Probabilistic Fracture Mechanics Model to the Influence of In-Service Inspection on Structural Reliability,"** *Probabilistic Fracture Mechanics and Fatigue Methods: Applications for Structural Design and Maintenance, ASTM STP 798,* J. M. Bloom and J. C. Ekvall, Eds., American Society for Testing and Materials, 1983, pp. 19–41.

ABSTRACT: A probabilistic fracture mechanics model of structural reliability is described that considers failure to occur as the result of subcritical and catastrophic growth of pre-existing cracks that escape detection. The model considers cracks to be two-dimensional and is capable of treating many of the input parameters as random variables and can consider arbitrary inspection schedules. The two-dimensional model is greatly simplified when one-dimensional cracks are considered, and an analytical treatment of the influence of in-service inspection for the one-dimensional case reveals that the ratio of failure rates with and without inspection is independent of the crack size distribution. Numerical results for two-dimensional cracks in a weld joint in a large reactor pipe show that the ratio of failure rates is not highly dependent on the initial crack distribution, even for this more general case. Thus, it appears that an assessment of the relative benefit of in-service inspection does not require accurate knowledge of the initial crack distribution. Additionally, the results show that leaks in large pipes are not very probable, but are much more likely to occur than a sudden double-ended pipe break.

KEY WORDS: probabilistic fracture mechanics, crack growth, in-service inspection, structural reliability, fatigue (materials), fatigue crack growth, fracture mechanics

Nomenclature

a Crack depth
a_c Critical crack depth
$a_k(t)$ depth of $a_{\text{tol}}(t)$ at times t_k

[1] Division managers, Science Applications, Inc., Palo Alto, Calif. 94304.

$a_{tol}(t)$	Tolerable crack depth—the depth of a crack at $t = 0$ that would grow to a_c in time t
a^*	Crack depth having a 50 percent chance of being detected during inspection
A	Parameter in nondetection probability
A^*	Parameter in nondetection probability
A_p	Cross-sectional area of pipe
b	Half surface-length of semi-elliptical crack
C	Parameter in fatigue crack growth relationship
C_β	Parameter in initial distribution of aspect ratio
D_b	Beam diameter of ultrasonic inspection
h	Thickness of pipe wall
H	$1 - \tfrac{1}{2}\,\text{erfc}\left(\dfrac{1}{\mu 2^{1/2}}\ \ell n\ \dfrac{h}{\lambda}\right)$
K	Stress intensity factor
ΔK	Cyclic stress intensity factor ($K_{max} - K_{min}$)
n	Number of fatigue cycles
$p_0(a)$	Probability density function of as-fabricated crack depths given that a crack is initially present
$p_0'(a)$	Probability density function of crack depth following a pre-service inspection given that a crack is initially present
$p_k(a)$	Probability density of crack depth just before the kth inspection given that a crack is initially present
$p_k'(a)$	Probability density of crack depth just after the kth inspection given that a crack is initially present
$P_0(a > x)$	Complementary cumulative distribution of as-fabricated crack depth given that a crack is initially present
$P_0'(a > x)$	Complementary cumulative distribution of crack depth following pre-service examination given that a crack is initially present
p_F	Failure rate (probability of failure per unit time)
p^*	Probability of initially having a crack in a body of volume V
p_v^*	Mean density of cracks
$P_{ND}(A)$	Probability of nondetection of a crack during inspection
R	K_{min}/K_{max}
t	Time
t_k	Time of kth inspection
t_F	Time to failure
V	Volume of the body considered
β	Equals b/a
β_m	Parameter in distribution of β
λ	Parameter in distribution of β (or a)
μ	Parameter in distribution of a
ν	Parameter in nondetection probability
σ	Normal stress
σ_{flo}	Flow stress of material

The purpose of this paper is to apply probabilistic fracture mechanics to the analysis of the influence of in-service inspection on structural reliability. The increased need for high performance or very high degrees of reliability or both have led to an increased interest in probabilistic analysis of structural reliability. Additionally, these considerations have resulted in increased reliance on in-service inspection as a means of detecting defects before they grow to a troublesome size. Probabilistic fracture mechanics provides a technique for estimating the probability of failure of a structure or component when such failures are considered to occur as the result of the subcritical and catastrophic growth of an initial crack-like defect. Such techniques are inherently capable of treating the influence of nondestructive inspections. For these reasons, probabilistic fracture mechanics is becoming of increased usefulness in analysis of the reliability of modern structures.

Attention will be concentrated in this paper on the influence of in-service inspection on reliability. Hence, this paper is a direct extension of earlier work [1,2],[2] but provides a considerable advance in that a general subcritical crack growth characteristic is considered (rather than just fatigue) and the results are presented in a more general form. Additionally, the procedures are expanded to consider more complex and realistic crack geometries that greatly complicates both the fracture mechanics and statistical consideration.

The earliest work in probabilistic fracture mechanics appears to have been related to aircraft applications [3-5], although general early discussions are available [6,7]. Becher and Pederson [8] provide one of the earliest applications to pressurized components in commercial power reactors, which is the area in which most recent efforts have been concentrated. Such applications are too numerous to review here, and Refs 9 through 11 provide additional information in this regard.

Probabilistic Fracture Mechanics Model

Probabilistic fracture mechanics models are generally based on the assumption that failure occurs due to the subcritical and catastrophic growth of crack-like defects introduced during fabrication. Such defects are initially present with a given probability, and are found during pre- and in-service inspections with a probability that is dependent on their size. The subcritical and catastrophic growth of these defects is governed by fracture mechanics considerations, which may also involve material properties that are randomly distributed. Cracks that are found by inspection are considered to be removed without introducing any new cracks into the material.

The majority of past work in probabilistic fracture mechanics has considered cracks to be one-dimensional, that is, the crack "size" can be expressed in terms of a single length parameter. Examples of such cracks are a single-

---

[2] The italic numbers in brackets refer to the list of references appended to this paper.

edge-cracked strip [1,2], a complete circumferential crack [12], or a semi-elliptical surface crack of fixed surface length-to-depth ratio [13]. A two-dimensional crack is much more realistic but considerably more complex. An example of a two-dimensional crack is a semi-elliptical surface crack of arbitrary surface length-to-depth ratio. A general probabilistic fracture mechanics model that is capable of treating two-dimensional cracks will be described in the following section. This will then be simplified to one dimension, and a treatment of the influence of in-service inspection in this simple case will be presented. The results of numerical calculations using the two-dimensional model will then be presented, and the degree to which they follow the trends predicted from the simple one-dimensional analysis will be assessed.

Description of General Model

The components of a probabilistic fracture mechanics model of structural reliability that considers the realistic case of two-dimensional cracks are presented in Fig. 1, which also shows the interrelationship of the various components. The approach is applicable to a wide variety of two-dimensional cracks, but the case of semi-elliptical surface cracks of arbitrary aspect ratio in a body of finite thickness will be considered here. Such a crack is shown schematically in the upper left corner of Fig. 1, and is characterized by two dimensions—a and b. The model depicted in Fig. 1 is described in detail in Refs *10* and *14*, so only a brief review will be presented here.

The procedures shown in Fig. 1 are applicable to a given location in a structure, such as a weld of volume V. The as-fabricated crack size distribution is combined with the nondetection probability to provide the post-inspection distribution. The manner in which the cracks that escape detection grow is then calculated by fracture mechanics techniques. The cumulative probability of failure at any time is simply the probability of having a crack at that time equal to or larger than the critical crack size.

The crack size distribution at the time of the first in-service inspection (ISI) can be calculated. This pre-inspection distribution is combined with the nondetection probability to provide the post-inspection distribution. Fracture mechanics calculations then proceed up to the next ISI, at which time the procedures are again applied. Calculations of the failure probability for the general model are performed numerically because of the complexity of the fracture mechanics calculations of the growth of two-dimensional cracks as well as the complicated bivariate nature of the crack size distribution.

A specific example of results from the general model will be presented later along with a discussion of inputs to the model—such as the nondetection probabilities, as-fabricated crack size distribution, and fracture mechanics material properties. Prior to this, the following discussion of a simpler one-dimensional model will be presented with specific emphasis on the influence of ISI.

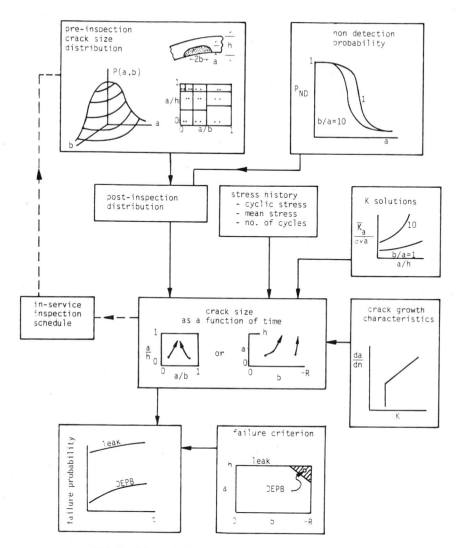

FIG. 1—*Schematic diagram of steps in analysis of reliability.*

One-Dimensional Model

Consideration of one-dimensional cracks greatly simplifies the probabilistic fracture mechanics model. The two-dimensional surface of the initial crack size distribution shown in Fig. 1 becomes a line that is a function of only one length, which will be taken to be crack depth a. Additionally, the fracture mechanics calculations are considerably simplified, because only one dimension of crack growth has to be considered. However, if one-dimensional cracks in finite bodies are considered, numerical calculations of subcritical crack growth are often still necessary.

The procedures involved for one-dimensional cracks are depicted in Fig. 2. For simplicity, only the case of a pre-service inspection is shown in Fig. 2, and the critical crack depth, a_c, is considered to be a constant.

Let $p_0(a)$ be the probability density function of as-fabricated crack depth—given that a crack is present. The as-fabricated complementary cumulative crack depth distribution is then given by

$$P_0(a > x) = \int_x^h p_0(y) dy \tag{1}$$

The density function and complementary cumulative distribution of crack depths following an inspection with nondetection probability $P_{ND}(a)$ are then given by

$$p_0'(a) = p_0(a) P_{ND}(a)$$
$$P_0'(a > x) = \int_x^h p_0'(y) dy = \int_x^h p_0(y) P_{ND}(y) dy \tag{2}$$

The probability of failure at short time following the pre-service inspection is simply the probability of having a crack of depth greater than the critical value, a_c. Hence, for $t \sim 0$, the Point A in Fig. 2 gives the cumulative failure probability—given that a crack is initially present.

During succeeding time, the cracks that are initially present can grow due to subcritical crack growth, such as fatigue or stress corrosion cracking. Fracture mechanics calculations could be performed to determine the entire crack size distribution as a function of time, such as shown by the dotted line in Fig. 2. The cumulative failure probability at or before t_s is then given by Point B in Fig. 2. Such a procedure requires extensive fracture mechanics calculations of crack growth in order to accurately define the crack size distribution at various times. An alternative procedure is to define $a_{tol}(t)$ to be the size of a crack at $t = 0$ (initially) that will just grow to a_c in time t—Fig. 2 schematically shows $a_{tol}(t)$. The probability of failure at or before t is then simply the probability of initially having a crack bigger than $a_{tol}(t)$. That is

$$P(t_F < t) = \int_{a_{tol}(t)}^h p_0(y) P_{ND}(y) dy \tag{3}$$

Hence, once $a_{tol}(t)$ is known, the cumulative (conditional) failure probability is calculated in a straightforward manner. The failure rate, or hazard

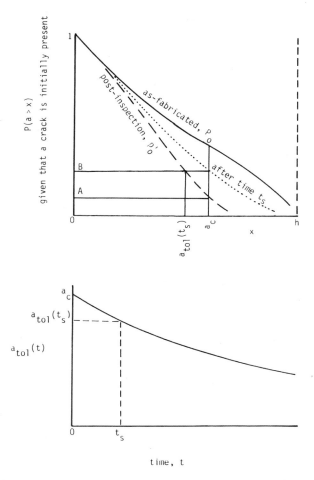

FIG. 2—*Schematic representation of procedures involved in calculating failure probability for one-dimensional crack problems.*

function, is equal to $(dP/dt)/(1 - P)$. In the case where $P(t_F < t) \ll 1$, the $1 - P$ term is nearly unity, and the failure rate is approximately given by the following

$$p_F(t) = \frac{d}{dt}[P(t_F < t)] = -\frac{da_{\text{tol}}(t)}{dt} p_0(a_{\text{tol}}) P_{ND}(a_{\text{tol}}) \quad (4)$$

Consideration of the failure rate in conjunction with the tolerable crack depth simplifies analysis of the influence of ISI. The procedures for calculating the failure rate for an arbitrary schedule of in-service inspections are straightforward but lengthy. Details are provided in the Appendix, which is drawn from Ref 12. The end result of the analysis is the following expression

$$\frac{p_F(t) \text{ with inspection}}{p_F(t) \text{ with no inspection}} = \prod_{n=0}^{k} P_{ND}[a_n(t)] \qquad (5)$$

The crack depth, $a_n(t)$, is shown schematically in Fig. 10 and is determined from fracture mechanics calculations. If $a_{\text{tol}}(t)$ is the depth of an initial crack that would just grow to a_c in time t, then a_n is the depth this crack would have had at the arbitrary inspection times t_n.

The result expressed as Eq 5 shows that the relative benefit of ISI, when expressed as the ratio of failure rates with and without inspection, is independent of the initial crack size distribution. This ratio is dependent only on the nondetection probabilities and crack growth characteristics. The initial crack size distribution is currently ill-defined, and is a source of considerable uncertainty in probabilistic fracture mechanics analysis.

Another ill-defined parameter is the probability of initially having a crack in a body of volume V. This parameter is denoted as p^*. All failure probabilities just discussed were conditional on having an initial crack. To account for the possibility of not having a crack (that is, to make the results *not* conditional on having a crack to begin with) all those results are to be multiplied by p^*. This parameter is also not well defined, but cancels out in the *relative* benefit of ISI as expressed by Eq 5, because it appears in both the numerator and denominator.

Thus, the *relative* benefit of ISI is independent of both the conditional initial crack size distribution and the probability of having a crack. These two factors are currently sources of considerable uncertainty, and their absence from the relative benefit allows the influence of ISI to be assessed with greater confidence then if they were present in Eq 5. This result is for only one-dimensional cracks. The degree to which it is applicable to the more realistic case of two-dimensional cracks will be addressed in the following section.

Results and Discussion

Results generated by use of the two-dimensional probabilistic fracture mechanics model depicted in Fig. 1 will be presented. These results were obtained by use of the PRAISE code described in Ref 15. This code utilizes stratified Monte Carlo techniques to handle the calculations involved in consideration of a bivariate crack size distribution that changes in time due to complex stress histories with material properties that are randomly distributed. References 10, 14, and 15 provide details, so only a brief summary of input parameters will be included here.

The problem to be considered is a weld joint in the primary cooling loop of a pressurized water reactor. The inside diameter pipe is 0.736 m (14.5 in.), with a wall thickness, h, of 6.35 cm (2.5 in.). The material considered is centrifugally cast austenitic stainless steel. The weld joint is subjected to the following axial stress history, which is considered to consist only of heatup-cooldown transients.

$$\sigma_{max} = 103.9 \text{ MPa (15.07 ksi)}$$

$$\sigma_{min} = 14.3 \text{ MPa (2.08 ksi)}$$

(load controlled stress at operating temperature)
$$= \sigma_{LC} = 59.1 \text{ MPa (8.57 ksi)}$$

This transient is taken to occur five times per year during the 40 year plant lifetime.

Initial Crack Size Distribution

The crack depth, a, and aspect ratio, $\beta = b/a$, are assumed to be independent. Two marginal distributions of crack depth will be considered. These density functions of crack depth are "corrected" to account for the impossibility of having cracks deeper than the thickness of the body, h.
Exponential (Marshall [13])

$$p_0(a) = \frac{e^{-a/\mu}}{\mu(1 - e^{-h/\mu})} \tag{6}$$

where $\mu = 6.25$ mm (0.246 in.).
Lognormal (Becher and Hansen [16])

$$p_0(a) = \frac{1}{\mu a H(2\pi)^{1/2}} e^{-\frac{1}{2\mu^2}(\ell n\, a/\lambda)^2} \tag{7}$$

where
$$H = 1 - \tfrac{1}{2} \text{ erfc}\left(\frac{1}{\mu 2^{1/2}} \ell n \frac{h}{\lambda}\right),$$

$\mu = 0.82$, and
$\lambda = 1.3$ mm (0.050 in.).

The complementary cumulative marginal distributions of crack depths corresponding to these density functions are presented in Fig. 3, which shows that the Becher and Hansen distribution is roughly two orders of magnitude below the Marshall distribution for cracks exceeding about 1 cm (0.4 in.) in depth.

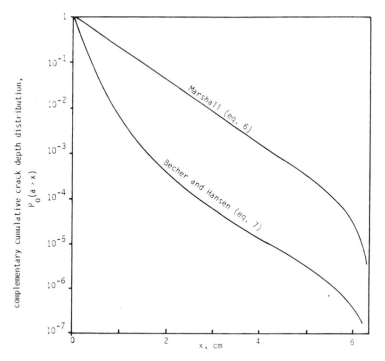

FIG. 3—*Complementary cumulative distributions of as-fabricated crack depth used in analysis. These results are conditional on a crack being initially present.*

The distribution of aspect ratio, β, will be taken to be as follows

$$p_0(\beta) = \begin{cases} 0 & \beta < 1 \\ \dfrac{C_\beta}{\lambda \beta \, (2\pi)^{1/2}} \, e^{-\frac{1}{2\lambda^2}(\ln \beta/\beta_m)^2} & \beta \geq 1 \end{cases} \quad (8)$$

where

$\beta_m = 1.336$,
$C_\beta = 1.419$, and
$\lambda = 0.5382$.

The number of cracks in a body of volume V will be assumed to be Poisson distributed with a mean density of $p_v^* = 10^{-4}/\text{in}^3$. The probability of having one or more cracks is denoted as p*, and is given by

$$p^* = 1 - e^{-Vp_v^*} \quad (9)$$

Nondetection Probability

Cracks in welded centrifugally cast austenitic stainless steel are difficult to detect. Therefore, the nondetection probabilities are relatively high. The following functional relationships for $P_{ND}(a)$ for an ultrasonic inspection is used.

$$P_{ND}(A) = \tfrac{1}{2} \, \text{erfc} \left(\nu \, \ell n \, \frac{A}{A^*} \right) \tag{10}$$

Figure 4 presents a plot of this relationship and defines the parameters in the equation.

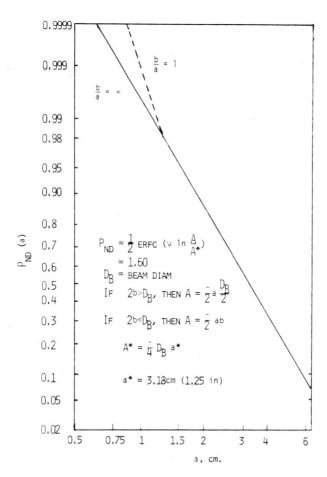

FIG. 4—*Two-dimensional lognormal model for probability of non-detection of a crack in cast austenitic material.*

Fracture Mechanics Characteristics

The crack growth law is taken to be given by the following

$$\frac{da}{dn} = C \left[\frac{\Delta K}{(1-R)^{1/2}} \right]^4 \quad (11)$$

where $R = K_{min}/K_{max}$, and $\Delta K = K_{max} - K_{min}$.

In order to account for the variability in the crack growth characteristics, parameter C is taken to be lognormally distributed with median 1.59×10^{-13} and standard deviation 3.83×10^{-13} (da/dn in metres/cycle, ΔK in MPa $- m^{1/2}$). [Corresponding values are 9.14×10^{-12} and 2.20×10^{-11} for da/dn in inches/cycle and ΔK in ksi·in.$^{1/2}$]. These results are based on experimental data [10].

Failure in a girth butt weld is considered to occur when a through-wall crack exists. Such a crack can result in a leak, or a sudden and complete double-ended pipe break (DEPB). Since the consequences of these two failure modes are very different, the probability of each occurring is of interest. Cracks are considered to grow as fatigue cracks until a leak develops, or until a DEPB occurs due to exceedance of a critical net section stress. Once a leak develops, it will either result in a DEPB or continue to grow as a through-wall fatigue crack. The critical net section stress criterion is expressed as

$$\sigma_{LC} A_p = \sigma_{flo} (A_p - A_{crack}) \quad (12)$$

where A_p is the pipe-cross-sectional area. The σ_{flo} is taken to be normally distributed with mean and standard deviation of 310 MPa (44.9 ksi) and 13 MPa (1.9 ksi).

Crack growth in the a and b directions is considered to be governed by Eq 11 with the appropriate value of ΔK. RMS-averaged values associated with each growth direction are employed [10,17]. Values of K for part-circumferential cracks in pipes are presented in Refs 10 and 18.

Results

Leak and DEPB probabilities as a function of time were generated for the set of preceding conditions. Results for Marshall and Becher and Hansen marginal depth distribution were obtained. An initial proof test to 1.25 times the design pressure was considered. The following three in-service inspection schedules were considered.

pre-service only $\quad t_k = 0$
inspection at 20 years $\quad t_k = 0, 20$
inspection every 10 years $\quad t_k = 0, 10, 20, 30$

The cumulative conditional leak and DEPB probabilities are presented in Figs. 5 and 6. These figures show that ISI does not generally have a large influence on the cumulative failure probabilities. This is undoubtedly due to the relatively high values of P_{ND} used here. The approximately two order of magnitude difference in the complementary cumulative crack depth distributions considered (see Fig. 3) is also observed in the two sets of leak and DEPB probabilities presented in Figs. 5 and 6. Additionally, there are about five orders of magnitude difference between leak and DEPB probabilities. The Becher and Hansen DEPB probabilities are very small, and are included here only in the interest of assessing the influence of $p_0(a)$ on the calculated failure probabilities. The effect is seen to be large, which is expected.

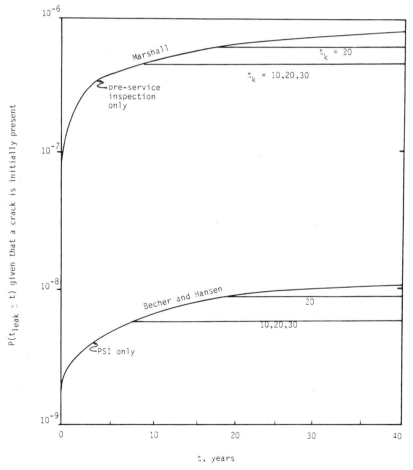

FIG. 5—*Cumulative conditional leak probabilities for various crack depth distributions and inspection schedules for example problem.*

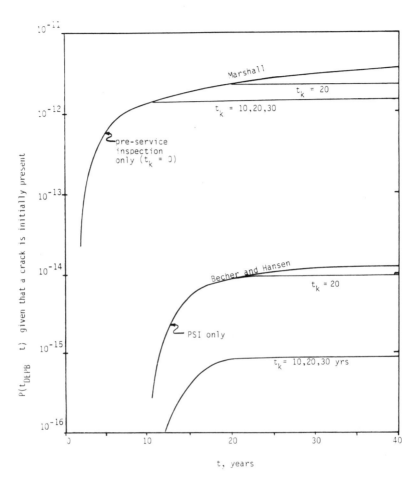

FIG. 6—*Cumulative conditional probability of a double-ended pipe break (DEPB) for various crack depth distributions and inspection schedules for example problem.*

The cumulative results summarized in Figs. 5 and 6 can be converted to failure rates, and the relative benefit of ISI determined. Figure 7 summarizes such results for leaks and DEPBs for both the Marshall and Becher and Hansen depth distributions. Results are presented for ISI divided by corresponding results for pre-service inspections only. Hence, the ratio is unity up to the first inspection that occurs at $t > 0$. In spite of the large differences in Figs. 5 and 6, the *relative* benefit of ISI is seen to be virtually independent of the initial depth distribution. Furthermore, the results are not highly dependent on the failure mode (leak versus DEPB), but inspections are somewhat more influential on leaks than DEPBs.

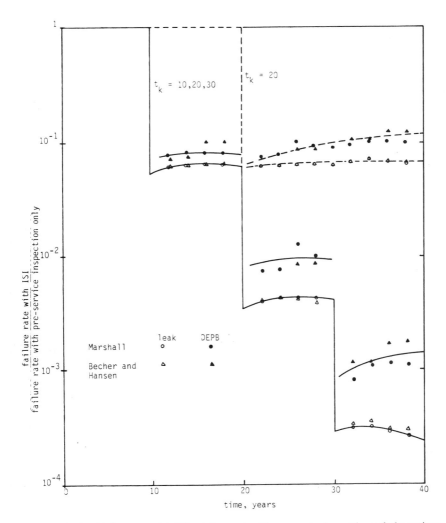

FIG. 7—*Ratio of failure rate with ISI to failure rate with pre-service inspection only for various marginal crack depth distributions, inspection schedules, and failure modes.*

It appears that the simple results (Eq 5) for one-dimensional cracks is carried over to the more complex and realistic case of two-dimensional cracks, and is not altered when the material properties are taken to be randomly distributed. However, the observation should not be considered to be totally general. There are probably instances where it would not be applicable, such as if P_{ND} was a stronger function of b/a than that shown in Fig. 4.

Summary and Conclusions

A probabilistic fracture mechanics model of structural reliability is summarized that considers cracks to be two-dimensional—such as semi-elliptical surface cracks. This model can treat the influence of in-service inspection but requires numerical techniques. The two-dimensional model is specialized to one-dimensional cracks, and a mathematical analysis of the influence of in-service inspections (ISI) presented for the case where the material properties are deterministic parameters. The results show that the relative benefit of ISI, when expressed as the ratio of failure rates with and without ISI, is independent of the initial crack size distribution—depending only on the detection probabilities and fracture mechanics results. Hence, the relative benefit of ISI can be assessed without requiring information on the initial crack size distribution.

Numerical results obtained for a weld in a large reactor pipe by use of the two-dimensional model are then presented for randomly distributed material properties and for two different initial crack depth distributions and three ISI schedules. The calculated cumulative failure probabilities are strongly dependent on the initial depth distribution, and always show a large difference between the leak and double-ended pipe break (DEPB) probabilities. Thus, it is predicted that a leak-before-break is much more likely than the reverse situation. ISI was seen to generally not have a large influence on the cumulative failure probabilities.

The leak and DEPB failure rates are obtainable from the numerical results and are cast in terms of the ratio of failure rates with and without ISI. The results show this measure of the relative benefit of ISI is virtually independent of the initial crack size distribution and also is not strongly dependent on the failure mode considered. Thus, even in the more general case it appears that the benefit of ISI can be assessed without detailed knowledge of the initial crack size distribution. However, the generality of the observations remains to be proven.

Acknowledgments

It is a pleasure to acknowledge the support for this work provided by the Electric Power Research Institute, Palo Alto, Calif., and Lawrence Livermore National Laboratory as part of their Load Combination Program.

References

[1] Harris, D. O., *Materials Evaluation*, Vol. 35, No. 7, July 1977, pp. 57-64.
[2] Harris, D. O., *Journal of Pressure Vessel Technology*, Vol. 100, No. 7, May 1978, pp. 105-157.
[3] Shinozuka, M. and Yang, J. H., *Annals of Assurance Science, Proceedings of the Reliability and Maintainability Conference*, Vol. 8, July 1969, pp. 375-391.
[4] Heer, E. and Yang, J. N., *Journal*, American Institute of Aeronautics and Astronautics, Vol. 9, April 1971, pp. 621-628.

[5] Yang, J. N. and Trapp, W. J., *Journal*, American Institute of Aeronautics and Astronautics, Vol. 12, Dec. 1974, pp. 1623-1630.
[6] Graham, T. W. and Tetelman, A. S., "The Use of Crack Size Distribution and Crack Detection for Determining the Probability of Fatigue Failure," AIAA Paper No. 74-393, presented at AIAA/ASME/SAE 15th Structures, Structural Dynamics and Materials Conference, Las Vegas, Nev., April 1974.
[7] Cramond, J. R., Jr., "A Probabilistic Analysis of Structural Reliability Against Fatigue and Fracture," Ph.D. thesis, University of Illinois at Urbana-Champaign, 1974.
[8] Becher, P. E., and Pedersen, A., *Nuclear Engineering and Design*, Vol. 27, No. 3, 1974, pp. 413-425.
[9] *Structural Integrity Technology*, J. P. Gallagher and T. W. Crooker, Eds., American Society of Mechanical Engineers, New York, 1979.
[10] Harris, D. O., Lim, E. Y., and Dedhia, D. D., "Probability of Pipe Fracture in the Primary Coolant Loop of a PWR, Vol. 5: Probabilistic Fracture Mechanics Analysis," Report NUREG/CR 2189, U.S. Nuclear Regulatory Commission, Washington, D.C., 1981.
[11] Johnston, G. O., *Welding Institute Research Bulletin*, Vol. 19, March 1978, pp. 78-82.
[12] Harris, D. O., "The Influence of Crack Growth Kinetics and Inspection on the Integrity of Sensitized BWR Piping Welds," Report EPRI NP-1163, Electric Power Research Institute, Palo Alto, Calif., 1979.
[13] "An Assessment of the Integrity of PWR Pressure Vessels," report of a study group chaired by W. Marshall, available from H. M. Stationery Office, London, England, 1976.
[14] Harris, D. O., and Lim, E. Y., "Applications of a Fracture Mechanics Model of Structural Reliability to the Effects of Seismic Events on Reactor Piping," *Progress in Nuclear Energy*, Vol. 10, No. 1, 1982, pp. 125-159.
[15] Lim, E. Y., "Probability of Pipe Fracture in the Primary Coolant Loop of a PWR, Vol. 9: PRAISE Computer Code User's Manual," Report NUREG/CR 2189, U.S. Nuclear Regulatory Commission, Washington, D.C., 1981.
[16] Becher, P. E., and Hansen, B., "Statistical Evaluation of Defects in Welds and Design Implications," Danish Welding Institute, Danish Atomic Energy Commission Research Establishment (no date).
[17] Cruse, T. A., and Besuner, P. M., *Journal of Aircraft*, Vol. 12, No. 4, April 1975, pp. 369-375.
[18] Lim, E. Y., Dedhia, D. D., and Harris, D. O., "Approximate Influence Functions for Part-Circumferential Interior Surface Cracks in Pipes," *Fracture Mechanism: Fourteenth Symposium, Volume I—Theory and Analysis*, 1983, pp. I-281-I-296.

APPENDIX

Analytical Procedures for Calculation of Effects of In-Service Inspection for One-Dimensional Cracks

The analytical procedures for calculating the effects of in-service inspection for one-dimensional cracks will be derived in this appendix, together with procedures for calculating the failure rates. As discussed in the text, the conditional cumulative failure probability at a given time (t) following a pre-service examination is simply the conditional cumulative probability of having a crack depth greater than $a_{tol}(t)$ following the pre-service examination. This will provide the failure probability up to the time of the first in-service inspection. The effects of the in-service examination on the crack depth distribution at that time must then be determined, which can be accomplished if the subcritical crack growth analyses have been performed. Such a sequence of calculations can be performed in order to determine the effects of an arbitrary number of in-service inspections.

Crack Depth Distribution

The following notation will be employed in order to simplify the bookkeeping involved in following the effects of successive inspections. Unless otherwise stated, all distributions are conditional on a crack being initially present.

$p_0(a)$ = initial as-fabricated crack depth distribution (p_0 of Eq 1)
$P_{ND}(a)$ = probability of *not* detecting a defect of depth a
$p_k(a)$ = pre-inspection distribution at the kth inspection
$p'_k(a)$ = post-inspection distribution at the kth inspection
t_k = time of the kth in-service inspection (t_0 corresponds to pre-service inspection, $t_0 = 0$)
a_k = crack depth at time t_k

The relationship between $p_k(a)$ and $p'_k(a)$ is the following

$$p'_k(a) = p_k(a) P_{ND}(a) \tag{13}$$

Consider the subcritical crack growth analysis to have been performed, with results such as shown in Fig. 8 being available. This figure simply presents the size of a crack as a function of time that was of size a_{ref} at $t = 0$. Time, t, or fatigue cycles could be used as the independent variable. Time will be used in this presentation. The results of Figure 8a can be replotted to provide Fig. 8b, which is applicable for crack growth over a given interval Δt. The results of Fig. 8b can be expressed as $y = G(x)$ or $x = G^{-1}(y)$.

Now consider the time, t_1, leading up to the first in-service inspection. The probability of having a defect with depth between $a_{\ell 0}$ and $a_{u 0}$ following the pre-service inspection is

$$\delta P = \int_{a_{\ell 0}}^{a_{u 0}} p'_0(x)dx \tag{14}$$

Let $a_{\ell 1}$ be the depth that the defect $a_{\ell 0}$ grows to in time t_1, and $a_{u 1}$ be the depth that $a_{u 0}$ grows to in this time interval. The following relationship then holds, using the definition $p_1(a)$.

$$\delta P = \int_{a_{\ell 0}}^{a_{u 0}} p'_0(x)dx = \int_{a_{\ell 1}}^{a_{u 1}} p_1(x)dx \tag{15}$$

The following relationships also hold from the definitions of Fig. 8b.

$$a_{\ell 1} = G(a_{\ell 0}) \qquad a_{\ell 0} = G^{-1}(a_{\ell 1})$$
$$a_{u 1} = G(a_{u 0}) \qquad a_{u 0} = G^{-1}(a_{u 1}) \tag{16}$$

Now, make the change of variable in the first integral of Eq 15, $x = G^{-1}(y)$ [or $y = G(x)$]

$$\delta P = \int_{G(a_{\ell 0})}^{G(a_{u 0})} p'_0[G^{-1}(y)] \frac{dx}{dy} dy = \int_{a_{\ell 1}}^{a_{u 1}} p_1(x)dx$$

$$= \int_{a_{\ell 1}}^{a_{u 1}} p'_0[G^{-1}(y)] \frac{d}{dy}[G^{-1}(y)] dy$$

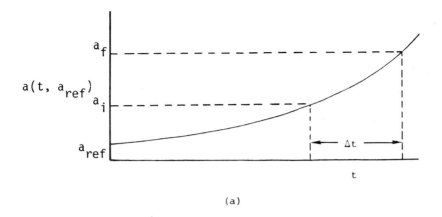
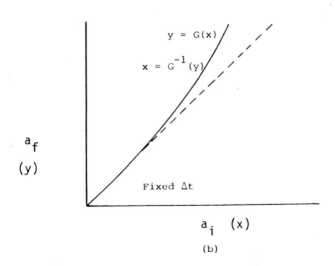

FIG. 8—*Schematic representation of results of subcritical crack growth analysis applicable to determination of crack depth distributions at times of in-service inspections.*

The following expression for $p_1(a)$ is therefore obtained

$$p_1(a) = p_0'[G^{-1}(a)] \frac{d}{da}[G^{-1}(a)] \qquad (17)$$

This is applicable at t_1. In order to remember this, let $a = a_1$, with a_1 being arbitrary. Let a_0 be the value a_1 had at $t = 0$. Then $G^{-1}(a_1) = a_0$, and the following expression results.

$$p_1(a_1) = p_0'(a_0) \frac{da_0}{da_1} \qquad (18)$$

Applying Eq 13 provides the following expression for the crack depth distribution following the first in-service inspection

$$p_1'(a_1) = p_0'(a_0) \frac{da_0}{da_1} P_{ND}(a_1)$$

$$= p_0(a_0) \frac{da_0}{da_1} P_{ND}(a_0) P_{ND}(a_1) \qquad (19)$$

Now consider the time between the first inspection (at t_1) and the second inspection (at t_2)

$$\delta P = \int_{a_{\ell 1}}^{a_{u1}} p_1'(x) dx = \int_{a_{\ell 2}}^{a_{u2}} p_2(x) dx$$

$$a_{\ell 2} = G_2(a_{\ell 1}) \qquad a_{\ell 1} = G_2^{-1}(a_{\ell 2})$$

$$a_{u2} = G_2(a_{u1}) \qquad a_{u1} = G_2^{-1}(a_{u2})$$

The G_2 corresponds to the function G in Fig. 8b but is now for $\Delta t = t_2 - t_1$.

Making the change of variable $y = G_2(x)$, $x = G_2^{-1}(y)$ in the first integral immediately preceding.

$$\delta P = \int_{G_2(a_{\ell 1})}^{G_2(a_{u1})} p_1'[G_2^{-1}(y)] \frac{dG_2^{-1}(y)}{dy} dy$$

$$= \int_{a_{\ell 2}}^{a_{u2}} p_2(y) dy$$

This provides the following result (where the value of a_2 is now arbitrary).

$$p_2(a_2) = p_1'[G_2^{-1}(a_2)] \frac{d G_2^{-1}(a_2)}{da_2}$$

By definition, $a_1 = G_2^{-1}(a_2)$, therefore

$$p_2(a_2) = p_1'(a_1) \frac{da_1}{da_2}$$

$$= p_0'(a_0) \frac{da_0}{da_1} \frac{da_1}{da_2} P_{ND}(a_1)$$

$$= p_0'(a_0) \frac{da_0}{da_2} P_{ND}(a_1) \qquad (20)$$

$$= p_0(a_0) \frac{da_0}{da_2} P_{ND}(a_0) P_{ND}(a_1)$$

The post-inspection crack depth distribution at t_2 is then given by

$$p_2'(a_2) = p_0'(a_0) \frac{da_0}{da_2} P_{ND}(a_1) P_{ND}(a_2)$$

$$= p_0(a_0) \frac{da_0}{da_2} P_{ND}(a_0) P_{ND}(a_1) P_{ND}(a_2) \qquad (21)$$

Generally, in going from the kth to the $(k + 1)^{\text{th}}$ inspection

$$\delta P = \int_{a_{\ell k}}^{a_{u k}} p'_k(x)dx = \int_{a_{\ell(k+1)}}^{a_{u(k+1)}} p_{(k+1)}(x) dx$$

$$a_{\ell(k+1)} = G_{k+1}[a_{\ell k}] \quad a_{\ell k} = G_{k+1}^{-1}[a_{\ell(k+1)}]$$

$$a_{u(k+1)} = G_{k+1}[a_{uk}] \quad a_{uk} = G_{k+1}^{-1}[a_{u(k+1)}]$$

Making the change of variable similar to that just employed, and following analogous steps lead immediately to the following result

$$p_{k+1}(a_{k+1}) = p'_k(a_k) \frac{da_k}{da_{k+1}}$$

The use of this result, along with the earlier ones for p_1 and p_2 provides the following general expression

$$p'_k(a_k) = p_0(a_0) \frac{da_0}{da_k} \prod_{n=0}^{k} P_{ND}(a_n) \tag{22}$$

This is the conditional crack depth distribution immediately following the kth inspection. For an arbitrary a_k, a_n is the value a_k had at the time of the nth inspection. Figure 9 schematically presents the means of determining a_n once a_k is fixed at an arbitrary selected value and the inspection times specified. The means of finding da_0/da_k is also shown. The results obtained in this appendix are for an arbitrary mechanism of subcritical crack growth—all that is needed is information such as shown in Fig. 8. These results are also applicable to an arbitrary inspection schedule.

The failure rates as a function of time can be evaluated now that the crack depth distribution following each inspection is known.

Failure Rates

The failure probabilities with in-service inspections can now be evaluated. Consider a time, δt, following the kth inspection. Let $P_F(\delta t)$ be the conditional cumulative probability that a crack with depth greater than $a_{\text{tol}}(\delta t)$ existed after the kth inspection, where $a_{\text{tol}}(\delta t)$ is the crack depth that would just grow to a_c in the time δt. P_F is then also the conditional cumulative probability of failure during the time δt following the kth inspection. From the definition of $p'_k(x)$, the following expression holds

$$P_F(\delta t) = \int_{a_{\text{tol}(\delta t)}}^{h} p'_k(x)dx \tag{23}$$

Let p_F be the conditional failure rate, then $p_F = dP_F/dt$

$$p_F = \frac{d}{dt} \int_{a_{\text{tol}(\delta t)}}^{h} p'_k(x)dx$$

$$= \frac{da_{\text{tol}}(\delta t)}{dt} \frac{d}{da_{\text{tol}}(\delta t)} \int_{a_{\text{tol}(\delta t)}}^{h} p'_k(x)dx \tag{24}$$

$$= -\frac{da_{\text{tol}}(\delta t)}{dt} p'_k[a_{\text{tol}}(\delta t)]$$

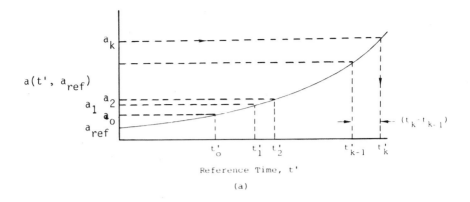

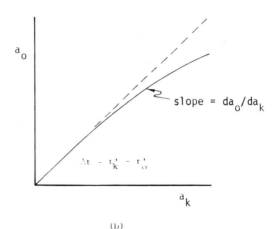

FIG. 9—*Schematic representation of procedures for determination of a a_n, once a_k is fixed and inspection times are specified. Also shown is the means of determining da_0/da_k.*

As shown schematically in Fig. 10, $a_{tol}(\delta t) = a_k$, therefore (using Eq 22)

$$\begin{aligned} p_F &= -\frac{da_k}{dt} p'_k(a_k) \\ &= -\frac{da_k}{dt} p_0(a_0) \frac{da_0}{da_k} \prod_{n=0}^{k} P_{ND}(a_n) \quad (25) \\ &= -\frac{da_0}{dt} p_0(a_0) \prod_{n=0}^{k} P_{ND}(a_n) \end{aligned}$$

By definition, $a_0 = a_{tol}(t)$ (t = total time), therefore

$$p_F = -\frac{da_{tol}}{dt} p_0(a_{tol}) \prod_{n=0}^{k} P_{ND}(a_n) \quad (26)$$

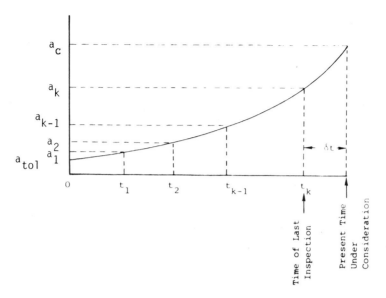

FIG. 10—*Schematic representation of determination of a_k from known subcritical crack growth results and specified inspection times.*

da_{tol}/dt is directly obtainable from the fracture mechanics analysis of subcritical crack growth.

The influence of the inspections can be easily visualized by dividing Eq 26 by the failure rate for no inspections. Denoting this as $p_{F(ni)}$, it can easily be evaluated from the initial crack depth distribution

$$p_{F(ni)} = \frac{d}{dt}\int_{a_{tol}(t)}^{h} p_0(x)dx = -\frac{da_{tol}}{dt}p_0(a_{tol}) \qquad (27)$$

Dividing this into Eq 26 provides the following remarkably simple result

$$\frac{p_F \text{ with inspection}}{p_F \text{ no inspection}} = \prod_{n=0}^{k} P_{ND}(a_n) \qquad (28)$$

Recall that a_n is the depth that a defect of depth a_{tol} had at t_n—the time of the nth inspection. Thus it is seen that the relative benefit of inspection, when expressed as in Eq 28, is independent of the initial crack depth distribution. The relative benefit of inspection depends only on the detection probabilities and subcritical crack growth characteristics of the system under consideration.

G. O. Johnston[1]

Statistical Scatter in Fracture Toughness and Fatigue Crack Growth Rate Data

REFERENCE: Johnston, G. O., "**Statistical Scatter in Fracture Toughness and Fatigue Crack Growth Rate Data,**" *Probabilistic Fracture Mechanics and Fatigue Methods: Applications for Structural Design and Maintenance, ASTM STP 798*, J. M. Bloom and J. C. Ekvall, Eds., American Society for Testing Materials, 1983, pp. 42–66.

ABSTRACT: The use of probabilistic fracture mechanics to assess the reliability of engineering structures requires knowledge about the statistical scatter of the parameters used in the models.
 A review is given of the statistically defined tanh reference toughness curves. This method is then used to aid an investigation into the effect of tighter chemical specifications. The predicted curves and small quantity of available published data do support the hypothesis of improved brittle/ductile transition temperatures with controlled chemistry. An experimental program has been undertaken to obtain plain plate data that are then used to derive an empirical correlation between K_{Ic} and Charpy energy, assuming the distributions of crack tip opening displacement and C_v can be related for lower shelf temperatures. Tolerance limits and sample size estimates are also obtained for specified confidence levels.
 Data on fatigue crack growth rates have been reviewed and the effects of parameters, such as stress ratio, temperature, and irradiation, studied qualitatively. A statistical analysis of published data has enabled a distribution for C in the Paris law to be derived assuming a linear $\ell n\, C$ versus m relationship.
 Finally, an illustrative example calculates the change in failure probability with the number of accumulated cycles.

KEY WORDS: fatigue (materials), steels, Charpy energy, fatigue crack growth, fracture toughness, fracture mechanics, probabilistic fracture mechanics, Paris law, tolerance limits

The fracture toughness of the material used in an engineering structure is a key input for any reliability assessment. Material properties tend to vary owing to inhomogeneity and nonuniformity of testing methods. Therefore, it

[1] Research officer, Berkeley Nuclear Laboratories, Central Electricity Generating Board, Berkeley, Gloucestershire, England.

is necessary to define acceptable limits rather than specify a deterministic value that must be achieved. The traditional method assumes that testing variability is negligible but that a curve may be used to describe the average behavior of the material under some external influence such as temperature changes. Available data are used to obtain the appropriate curve from which a lower bound design (or reference) curve can be defined. Statistical models have been derived in the United States to predict fracture toughness reference curves from Charpy data [1,2].[2] Using these methods, the effect of tightening the chemical specifications of a material has been predicted and comparisons made with available data, in the literature.

An allied problem is that of scatter observed in nominally identical fracture toughness tests. Statistical distributions may be used to take account of the variability, and an experimental program that has been designed statistically is described. The ultimate aim is to predict fracture toughness and the confidence that can be placed in a small sample of test results. This approach provides a valuable input for the calculation of the reliability of a structure in terms of the probability of safely meeting the design life. Many factors will contribute to this calculation. In the simplified case, two basic parameters govern the failure probability, namely, the defects present (for example, in a weldment) and the toughness of the material, which through fracture mechanics yields a critical crack size distribution. This assumes that other parameters, such as stress, are determinate. Failure then occurs when a defect is larger than the critical crack size.

The distribution of defect sizes poses a real problem since it is hard to obtain realistic data. Currently available information on this subject has been recently reviewed [3]. This distribution may change during the service life of the structure, for example, by repairing defects found during a nondestructive examination or by fatigue crack growth. Fatigue failure is the most common cause of structural failure and fatigue cracks pose a serious problem in design. In complex structures, fracture mechanics can be used to determine the influence of such cracks as they grow from defects situated in highly stressed regions. Fatigue crack growth rate (da/dN) expressed as a function of the range of stress intensity factor (ΔK) at the crack tip characterizes a material's resistance to crack extension under cyclic loading. A simple power law was proposed by Paris [4] to relate da/dN and ΔK, namely,

$$\frac{da}{dN} = C(\Delta K)^m \tag{1}$$

The scatter about this line can vary widely for experimental tests and will depend on the conditions of the test. Many data are available for a variety of conditions, investigated as part of the Heavy Section Steel Technology Pro-

---

[2]The italic numbers in brackets refer to the list of references appended to this paper.

gram (HSSTP) in the United States. These and data from other sources have been reviewed to indicate the possible effects of parameters such as stress ratio, temperature, neutron irradiation, and environment. In some cases, scatter confuses any possible variation, so that the qualitative conclusions drawn here should be followed up with statistical arguments.

One way of taking account of such scatter is to introduce distributional descriptions of C and m in the Paris law to replace the deterministic values. In reality, it seems likely that the required distributions may depend on the parameters just mentioned. However, there are generally insufficient data for one particular set of conditions to enable a statistical analysis. As a first approximation, the data may be pooled. Previous analyses by Gurney [5] for structural steels indicate a relationship between C and m. Assuming the statistical scatter of C is constant for each particular m, Gurney's data have been transformed to one value of m. This yielded sufficient data to derive a distribution for C.

Finally, a simple example has been used to demonstrate the calculation of a failure probability. Since the purpose is to give an indication of the method rather than to perform complicated numerical analysis, the assumptions may be thought to be convenient rather than the most realistic possible.

A Review of the Development of Tahn Fracture Toughness Reference Curves

Fracture toughness varies with temperature and a lower bound curve, the K_{IR} curve, based on a limited amount of data, is given in the ASME Boiler and Pressure Vessel Code, Section III, 1980. This represents the lower bound of static, dynamic, and crack arrest data from specimens of material to SA533B Class 1 and SA508 Classes 1, 2, and 3. Only a few heats of material were used, though the curve is commonly assumed to apply for all ferritic materials approved for nuclear pressure boundary applications having a minimum specified yield strength of 345 MPa or less, at room temperature.

The curve is shifted along the temperature axis according to the value of RT_{NDT}, an index reference temperature. This can be determined experimentally for a particular material by a few small-scale tests and hence the appropriate lower bound, K_{IR}, can be found for the temperature, T, of interest

$$K_{IR} = 29.43 + 1.34 \exp \{0.0261(T - RT_{NDT} + 89)\}$$

(temperature in degrees centigrade and K_{IR} in MPa$\sqrt{m}$).

However, RT_{NDT} provides a poor basis for describing heat-to-heat differences since it is rather an indirect measure of toughness and since the variance of the test is quite high [6]. It is unable to take account of the positions of the lower and upper shelves of the transition curve. The former was fixed at 30.8 MPa$\sqrt{m}$, while the latter was ignored so that the curve increased exponentially with temperature. Another disadvantage was the lack of statistical basis for the lower bound. The curve was generally found to be overconservative in the transition region, but the degree of conservatism was unpredictable for different heats of material.

The inadequacy of the ASME K_{IR} curve to predict heat-to-heat behavior indicated the need for further experimental work and associated statistical analyses. An extensive program was initiated to obtain fracture toughness data on 57 heats covering a range of materials—including A533B, A508, and associated weldments [7]. Fabrication details were recorded in each case as well as chemical and mechanical properties. Charpy results are available for each heat [8], which have subsequently been used to make predictions regarding the fracture toughness of larger scale specimens. A few experimental results are also given and these are found to agree well in most cases with the predicted curves. The method [2] is summarized since it will be used in a specific example later in the paper.

In order to characterize a complete Charpy energy or fracture toughness curve, four parameters are required to account for the brittle/ductile transition temperature, the lower and upper shelf behavior, and the rate of increase of toughness with temperature in the transition region. A convenient description [2] can be made in terms of the tanh curve, which defines toughness K, or Charpy energy, in terms of temperature, T, as

$$K = A + B \tahn \left(\frac{T - T_0}{C} \right)$$

where $A - B$ and $A + B$ define the lower and upper shelf toughness, respectively, T_0 is the temperature at the midpoint of the transition, and B/C is the slope in the transition region.

A procedure has been developed in the United States by Oldfield [2] for predicting tanh fracture toughness curves from similar shaped Charpy curves, assuming sufficient small-scale specimen results are available. Here, sufficiency means enough to produce a good estimate of the Charpy transition curve and hence the parameters a, b, t_0, and c of the tanh model

$$k = a + b \tanh \left(\frac{t - t_0}{c} \right)$$

where $k = \sqrt{EC_v}$ is chosen to remove the inconsistency in units between Charpy energy, C_v and K, with E, a temperature dependent elastic modulus.

From a, b, t_0, and c, a set of bias coefficients can be used to predict fracture toughness curves for the mean and also various statistical bounds. The theoretical work has been well reported in the literature [9]. In most cases, the available fracture toughness data did lie above the predicted lower bounds based on Charpy data from the same material. However, it must be emphasized that representative Charpy data are required, taking into consideration possible systematic variation of plate properties.

The Effect of More Rigid Chemical Specifications

Recently, there have been considerable improvements in the properties of reactor pressure vessel steels, which indicate a possible tightening of materials' specifications. The control of chemical composition may itself cause an improvement in fracture toughness. Few data are currently available on the "new" materials, but the initial results are encouraging.

The recommendations of the Marshall Committee [10] are that the levels of carbon, sulfur, and copper should be restricted to below 0.2, 0.015, and 0.1, percent, respectively, for the base material. For weld metal, these become 0.2, 0.010, and 0.05 percent. In the EPRI data base [8], only two heats satisfy these requirements for all three elements, namely, Heats G and H. The 1-in. compact tension data are plotted in Fig. 1 with data from other heats, and they can be seen to lie in the upper half of the scatter band. Other published data [11,12] are also consistent with this trend. Therefore, these results do not contradict the hypothesis of improved toughness, but so few hardly prove the desired drop in the brittle/ductile transition temperature.

A large program is currently in progress at Staatliche Materialprufungsanstalt (MPA) University of Stuttgart, and this may help clarify the effect of chemical composition on fracture toughness. An indication of the initial

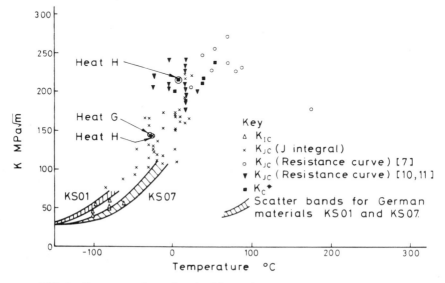

FIG. 1—*Fracture toughness data for 1-in. static compact tension tests on A533B.*

results has been presented in the form of scatterbands [13]. These bands are reproduced in Fig. 1, but can be seen to encompass only a small temperature range, around the brittle/ductile transition temperature. Nevertheless, the two bands are nonoverlapping representing an "average" temperature difference of about 30°C. The lower set of results (KS07) does correspond to a relaxation in the chemical composition as compared with KS01 that is essentially A508-2, but neither material falls within the tighter specifications proposed by the Marshall Committee. The effects of the various chemical elements present have been examined elsewhere in more detail, with reference to material properties and Charpy energy [14].

In the absence of full-scale data, an alternative approach is to compare fracture toughness curves predicted from Charpy data. In this way, it might be possible to estimate any improvement in toughness associated with the "new" materials. Heats D and G were chosen from the EPRI data base since the former represented an "old" material and the latter conformed to the tighter chemical specifications as shown in Table 1. The predicted curves using Oldfield's method are shown in Fig. 2. These assume a fixed lower shelf of 0.801 MPa$\sqrt{m}$ for the Charpy data. Since the RT_{NDT} for Heats D and G was -12 and -18°C, respectively, the ASME K_{IR} curve is also shown for comparison displaced by -15°C.

In each case, the mean curve and 95 percent tolerance bound on fracture toughness are given. There is a predicted temperature shift in the transition region between the two heats of material. This represents a shift of 10°C for the mean curve and 15°C for the 95 percent tolerance curve with the improved toughness corresponding to the tighter material specifications. In contrast, the "improved" material appears to exhibit an upper shelf value reduced by 15 MPa$\sqrt{m}$.

Statistical Scatter of Fracture Toughness Data

Ideally, it is desirable to know the scatter about these curves in more detail. As mentioned earlier, there are rarely sufficient data for a particular set of conditions to permit a statistical analysis. The only safe way to obtain such data is by setting up large experimental programs specifically designed to investigate this problem. Necessarily, such work will be limited to some degree by resources—both time and money available.

An initial program of work has been completed on plain plate steel to BS4360 Grade 50D. This comprised crack tip opening displacement (CTOD) measurements of toughness. A transition curve was obtained by testing a number of specimens at varying temperatures. The fracture surfaces were then examined to identify the different modes of failure and three temperatures were chosen to represent cleavage initiation and ductile initiation followed either by cleavage instability or by plastic collapse. At each appropriate temperature, namely, -100, -60, and 0°C, thirty CTOD tests

TABLE 1—*Chemical properties for A533B and A508 Cl 2 base material.*

	C	Mn	P	S	Si	Ni	Cr	Mo	V	Cu
				ASTM Specifications (maximum or range)						
A533B	0.25	1.07–1.62	0.035	0.04	0.13–0.32	0.37–0.73	...	0.41–0.64	...	...
A508-2	0.27	0.5 –1.0	0.025	0.025	0.15–0.40	0.5 –1.0	0.25 –0.45	0.55–0.70	0.05	...
				German Material (maximum)						
	C	Mn	P	S	Si	Ni	Cr	Mo	V	Cu
KS01	0.25	0.71	0.009	0.022	0.24	0.95	0.41	0.75	0.012	0.11
KS07	0.32	1.01	0.02	0.011	0.30	0.76	0.56	1.03	0.01	0.26
				EPRI A533B Material						
	C	Mn	P	S	Si	Ni	Cr	Mo	V	Cu
Heat D	0.24	1.51	0.008	0.009	0.30	0.48	0.02	0.54	0.003	0.13
Heat G	0.17	1.32	0.008	0.013	0.23	0.59	0.05	0.55	0.004	0.10

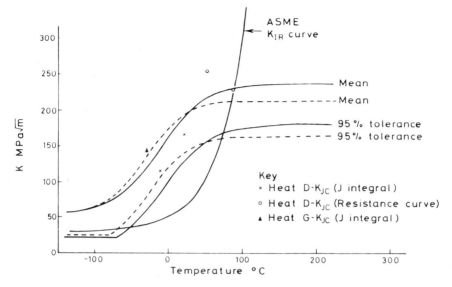

FIG. 2—*Fracture toughness predictions for A533B Heats D (——) and G (----)*.

were performed under nominally identical conditions. The specimens were extracted from the plate and, to allow a statistical analysis, they were assigned one of the three test temperatures at random.

All the CTOD values were calculated according to British Standard BS5762: 1979. The results are presented as histograms in Fig. 3. At $-60°C$, cleavage and maximum load values were obtained and these lay in the lower and upper regions of the scatter band, respectively, suggesting the generally postulated overlap of distributions (one for each failure mode). In each case, best fitting normal, lognormal, and Weibull distributions were derived using the method of moments. It was found that a lognormal model gave a reasonable fit to the data at $-100°C$ and likewise the Weibull distribution at -60 and $0°C$. The assumption of normality was unrealistic, predicting negative CTODs.

It is perhaps interesting to note that the new British Standard BS5672 for CTOD appears to give more conservative (that is, lower) values than the Wells' formulas of DD19: 1972. On average, a reduction of 10 percent in CTOD may be observed.

The study on plain plate toughness, which was just mentioned also included an investigation into the distribution of Charpy energies. The same experimental plan was followed, but with 50 specimens at each of three temperatures, representing cleavage, transition region behavior and void coalescence. The corresponding temperatures were -60, -20, and $+20°C$,

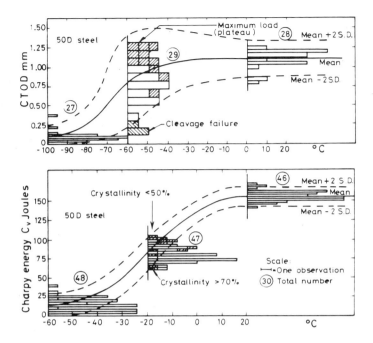

FIG. 3—*Summary of test results showing mean transition curves.*

and the results are included in Fig. 3. Best fitting distributions were found to be Weibull, lognormal, and lognormal, respectively.

Further analyses of the data were not included in the initial work so the aim of the following section is to make use of the experimental results to investigate a possible correlation between Charpy and K_{Ic} values.

Since Charpy tests are quick and easy to perform, there has been much interest in the possibility of fracture toughness K_{Ic} predictions from Charpy energy, C_v using empirical formulas [15]. To derive a relationship between K_{Ic} and C_v under linear elastic conditions, the following method may be applied. For simplicity, it is assumed that

$$\frac{K_{Ic}^2}{E} = \alpha C_v^\beta$$

where α and β are to be determined.

From BS5762, K_{Ic} and CTOD, δ, are related under linear elastic conditions by

$$\delta = \frac{K_{Ic}^2}{E} \cdot \frac{(1-\nu^2)}{2\sigma_y}$$

Therefore

$$\delta = \frac{(1-\nu^2)}{2\sigma_y} \cdot \alpha C_v{}^\beta \qquad (2)$$

α and β may be determined by considering the distributions of δ and C_v at $-100°C$ and $-60°C$, respectively, under the assumption that both temperatures represent lower shelf behavior. Equation 2 then becomes

$$\ln \delta = \ln \kappa + \beta \ln C_v$$

where for $\nu = 0.3$, and $\sigma_y = 430$ N/mm², it follows that $\kappa = 0.00106 \, \alpha$. Under the assumption of lognormality, the distribution of δ can be calculated from that of C_v and vice versa since $\ln \delta$ has a normal distribution if $\ln C_v$ is normal. Equating the means and variances of the best fit distributions will then give two equations that may be solved for α and β.

Thus, if

$$\ln \delta \cap N(-2.517, 0.504^2)$$

$$\ln C_v \cap N(2.44, 0.68^2)$$

then

$$-2.517 = \ln \alpha + \ln 0.00106 + \beta \cdot 2.44$$

$$0.504^2 = \beta^2 \, 0.68^2$$

So

$$\beta = 0.74$$

$$\alpha = 12.5$$

(As a check, note that substituting the average value for $C_v = 14.1$ J, gives $\delta = 0.094$ mm for the calculated values of α and β. This is in good agreement with the average $\delta = 0.093$ mm obtained experimentally.)

In this case, under the assumption of lognormality, and LEFM, K_{Ic} and C_v can be related according to the equation

$$\frac{K_{Ic}{}^2}{E} = 12.5 \, C_v{}^{0.74}$$

Closely allied with this is the problem concerning the prediction of fracture toughness from a small number of specimens. One approach is to calculate tolerance limits from sample data, that is a lower limit, above which a prescribed quantity of future data would be expected to lie with a specified confidence, say 90 percent. The distributions of CTOD are assumed to be lognormal. Tolerance bounds on $\bar{x} - ks$ may be calculated by considering confidence limits on $\bar{x}$ and s individually. The derivation follows the method of Johnson [16]. It was previously applied to the $\mu - 2\sigma$ limit in fatigue by Johnston [17]. Suppose that $(1 - \beta)$ 100 percent of future data is required to lie above tolerance limit, TL, with a confidence of $(1 - \alpha)$ 100 percent, then

$$TL = \bar{x} - ks - ds \qquad (3)$$

where

$$k = \left(1 + \frac{1}{n}\right)^{1/2} t_{1-\beta}$$

$$d = \frac{t_{1-\gamma}}{\sqrt{n}} + k\left(\sqrt{\frac{n-1}{\chi_\gamma^2}} - 1\right)$$

$$1 - \gamma = \sqrt{1 - \alpha}$$

The derivation of Eq 3 is similar but more general (in fatigue, $k = 2$ and $\beta = 0.025$). Though more complex formulas are available [18], the results in this case were not appreciably different from the simpler approach used here. Typical results are given in Table 2, for the plain plate CTOD data. In all cases, a 90 percent confidence has been assumed, with the percentage of data above the limit [$(1 - \beta)$ 100 percent] being 90, 80, or 50 percent. Note that the 50 percent tolerance limit provides a 90 percent confidence in the value above which 50 percent of the data will lie, that is, in the median value of CTOD because of lognormality. The median values are included in Table 2 for comparison. The larger difference in values at $-60°C$ just reflects the scatter in the transition region.

Rearranging, Eq 3 gives

$$t_{1-\beta} = \left(\frac{\bar{x} - TL}{s} - \frac{t_{1-\gamma}}{\sqrt{n}}\right)\left(\frac{n\chi_\gamma^2}{n^2 - 1}\right)^{1/2}$$

TABLE 2—*Tolerance limits.*

Temperature, °C	90%	80%	50%	Median CTOD, mm
−100	0.028	0.039	0.068	0.081
−60	0.22	0.31	0.58	0.70
0	0.85	0.91	1.04	1.08

so that β can be calculated from a knowledge of the other parameters. In particular, if TL is taken as the minimum value of CTOD from the plain plate tests, then the percentages of data lying above TL with 90 percent confidence are 92.0, 97.4, and 96.0 percent corresponding to -100, -60, and $0°C$, respectively.

These can now be compared with a theoretical prediction dependent only upon sample size and the required confidence [19]. No assumptions are made about the underlying distribution. In the preceding notation

$$(1 - \beta)^n = \alpha \tag{4}$$

where $(1 - \beta)$ 100 percent denotes the percentage of data lying above the minimum of n tests. Here, $1 - \alpha = 0.90$ and $(1 - \beta)$ 100 = 91.8, 92.4, and 92.1 percent, each of which is lower than the corresponding value above. This reflects the conservatism of the more approximate method.

For practical purposes, Eq 4 may be used to estimate the adequacy of a small number of tests, (that is, the sample size to give a required confidence in the minimum). For example, for three tests and a confidence of 90 percent, Eq 4 predicts that at least 46 percent of future data will lie above the minimum of the three. This rises to 68 percent by increasing the sample size to six. Therefore, this is indicative that care is required in interpreting series of three CTOD tests and that where more tests are possible, the confidence is greatly increased. Alternatively, Eq 3 may be used to estimate the sample size, but it does rely upon prior information of the distribution shape and this may not always be available.

Statistical analyses of systematic variation in fracture toughness (for example inter-laboratory differences) have been reviewed elsewhere [3].

Fatigue Crack Propagation

In this section, qualitative results of a literature survey and subsequent analysis of fatigue crack growth rate data are presented. The ultimate aim was to investigate statistically the characteristics of the variation of da/dN with ΔK. The initial review is presented here to illustrate the scatter inherent in fatigue crack propagation data. Full use of the results must await the completed statistical analysis. Since the completion of this work, further data have been published (see for example, Cullen et al [20]), but these are not included in the data reported here.

The conditions were chosen primarily with pressurized water reactor (PWR) pressure vessel applications in mind. Table 3 summarizes the references included in the analysis and indicates some of the materials and conditions covered by the results. All data are from constant amplitude fatigue tests carried out on one of five types of specimen, namely, wedge opening loading (WOL), compact tension (CT), center cracked tension

TABLE 3—Summary of conditions covered in each reference.

Refer-ence	Material			Environment		Irradiated		Stress Ratio							Temperature, °C					Specimen Thickness, in.					Frequency, Hz			Specimen Geometry
	A533B	A508	weld	water	air	yes	no	0	0.1	0.2	0.3	0.5	0.7	0.8	0	24	medium	288	other	<1	1	2	3	4	<1	1-10	160	
21	x	x	x	x			x			x								x	x			x			x			WOL
22	x	x					x		x	x	x	x	x			x	x	x	x								x	CT
23	x				x		x		x							x				x						x		CT
24	x				x		x		x							x					x					x		WOL
25	x		x		x	x	x									x								x	x			CCT
26	x						x			x								x						x	x			CT
27	x		x	x		x												x				x				x		WOL
28	x		x	x		x				x								x				x				x		WOL
29	x			x		x										x	x	x				x	x	x		x	x	WOL
30	x				x	x									x	x		x				x	x	x		x		WOL
31	x				x	x									x		x						x	x		x		WOL
32	x				x	x												x			x	x				x		{CT, SENB}
33	x				x		x											x			x	x			x	x		WOL
34	x				x	x										x				x	x				x			DEC
35		x			x	x										x				x	x				x			
36	x				x	x												x		x					x			SENB
37	x				x	x												x			x				x	x		CT

(CCT), single edge notch cantilever bend (SENB), or double edge crack (DEC). No account has been taken of possible variations caused by different geometries since expressing da/dN as a function of ΔK theoretically provides results that are independent of specimen geometry. Specimens were loaded in tension (nonnegative stress ratio) axially or in bending and loading frequency varied from 0.0017 Hz (0.1 cpm) to 160 Hz. Both PWR water and air environments were considered with temperatures ranging from 0 to 343°C. Most results were at room temperature (24°C) or 288°C. The effect of irradiation was also investigated. Fatigue crack propagation has been treated by linear elastic fracture mechanics and the data are presented irrespective of the stress condition (since it was not always clear whether or not the specimens failed in conditions of plane stress or plane strain).

Few raw data, that is crack length versus number of applied cycles, were available since most results were presented in the form of points on da/dN versus ΔK graphs. It was assumed that all techniques of calculating da/dN gave equivalent results, so that the da/dN and ΔK values were extracted straight from the graphs. Data points were taken using a flat-bed digitizer to measure the distances in millimetres. The data were collected on paper tape for direct input into the computer, which was then programmed to analyze the data and draw the appropriate graphs. This minimized the error in the whole process.

The data presentations are divided into six sections, each of which considers the variation of different parameters according to the availability of data. These six divisions investigate the possible effects of:

1. Material differences.
2. Weld metal.
3. Environment.
4. Neutron irradiation.
5. Specimen thickness and temperature.
6. Stress ratio and temperature.

In some cases, scatter confuses any possible variation, so that conclusions are impossible. This also occurs where several sets of data are pooled, as shown by the data in Fig. 4 [*21*].

Material Differences [22-24]

The only data available to provide a comparison between the typical PWR steels, namely, A533B and A508 are given by Paris et al [*22*]. A large quantity of data is reported so that comparisons from tests conducted under the same (nominal) conditions are possible. In particular, there are results for differing temperatures and stress ratios. In most cases, the scatter band for the A533B data encompassed the sparser A508 data. Alternatively, slightly higher fatigue crack growth rates were observed for A508 under certain con-

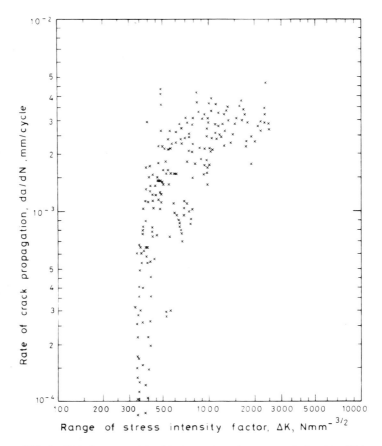

FIG. 4—*Data from a variety of tests causing a large amount of scatter [21].*

ditions, for example at 288°C and a stress ratio, R, of 0.1. Since in the present analysis, differences in parameters are of more interest, the A508 results have been excluded from most of the following work. This tends to reduce the scatter so that trends are more apparent.

Weld Metal [25-28]

Most of the data on weld metal were associated with PWR water environment. For base material (A533B), it was observed that higher crack growth rates were associated with a reduction of testing frequency from 1 to 0.017 Hz, whereas there was little difference between 1 and 10 Hz under a PWR water environment. Accordingly, the 1 and 10 Hz data were used as a comparison for the weld metal data. Figure 5 gives typical results, from which it is difficult to distingish any real trends.

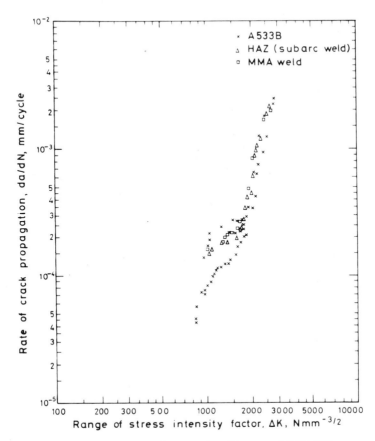

FIG. 5—*Comparison of base metal, weldment, and HAZ [25].*

Environment [21,26-30]

Only five of the references include data for PWR water environment and two of these [21,29] are unsuitable for investigating environmental effects, since the data for a variety of conditions are grouped together. The third [26] includes specimens tested at a low frequency that led to increases in crack growth rate, as shown earlier. From the data in the remaining papers [27,28], a large amount of scatter is apparent. For comparison, Mager and McLaughlin [28] do report some data in air for $R = 0.1$, under similar conditions, which tend to show a slight drop in fatigue crack growth rates between ΔK values of 2000 and 2400 Nmm$^{-3/2}$. It is perhaps interesting to note that the other "in air" data that they quote [30] was for a stress ratio of $R = 0$. These data show a much lower crack growth rate than the $R = 0.1$ data. It seems reasonable to assume that these lower values are due in part if not wholly to a stress ratio effect, which was observed for tests conducted in air under other conditions.

Neutron Irradiation [30-37]

The data obtained from specimens subjected to neutron irradiation seem to present a confused picture, even when different testing conditions are taken into account. Data have been obtained using a variety of testing frequencies and specimen geometries, at room temperature (24°C) and 288°C. Additional data on unirradiated material are drawn from tests under similar conditions. At 24°C, the irradiation data fall within the scatter band for unirradiated specimens, but at 288°C there is a more marked effect. Unlike the results in water mentioned earlier, the unirradiated data in air seem unaffected by a decrease in frequency. In contrast, irradiated data show an increase in crack growth rate associated with the lower frequency. This has the strange effect that for high frequencies, irradiation produces lower crack growth rates whereas the opposite is true for lower frequencies (at 288°C).

Specimen Thickness and Temperature [30-31,37]

There is a reasonable amount of data from 1T, 2T, 3T, and 4T WOL and 1T compact tension specimens. At room temperature, the scatter band for the 1-in. specimens encompasses all the other results. For a given specimen size, the scatter band width tends to increase with ΔK. Additionally, the overall spread is least for the larger specimen sizes since they are more representative of the material. This means that localized metallurgical variations will have less influence on the crack growth rate. At 288°C, there is an apparent difference in crack growth rates as shown in Fig. 6, but no consistent effect.

This confused pattern is also observed in consideration of temperature effects. For the 1-in. specimens the data at 288°C form an upper bound for the room temperature results. This is in direct opposition to the trend for the other specimen sizes and may conceivably be due to a specimen geometry effect. For the 2-in. specimens, the 24°C data provide the upper bound, and for 3-in. there is no distinguishable difference due to temperature. Typical results for 1-in. specimens tested at $R = 0$ are given in Fig. 7.

Stress Ratio and Temperature [22]

Returning to the results by Paris et al [22], the effect of stress ratio and temperature can be investigated. Here, one of the sets of A508 data was included since no A533B data were reported. This was at room temperature and $R = 0.1$. In this case, any lowering of propagation rate for A533B (as compared with A508) would increase the effect noticed since the lowest stress

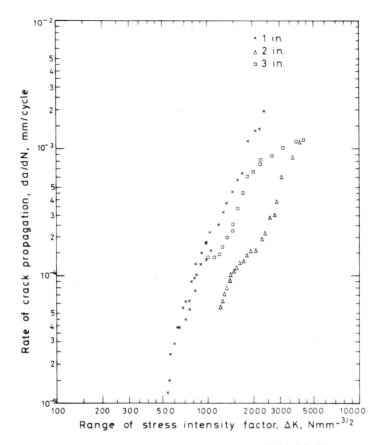

FIG. 6—*The effect of specimen thickness at 288°C [27,35].*

ratio ($R = 0.1$) gave the lowest fatigue crack growth rates. At room temperature, there is a marked fork in the data for low ΔK, which decreases as temperature increases until it is no longer visible at 288°C. Examination of Fig. 8 shows that the results for $R = 0.3$ and 0.5 may be grouped together and similarly for $R = 0.7$ and 0.8. At higher temperatures, the same effect is observed, though it is less apparent since the overall scatter band is narrower.

As R increases so does da/dN at 24°C and for low ΔK values. The apex of the fork is around 200 Nmm$^{-3/2}$ for the $R = 0.7/0.8$ results compared with the $R = 0.3/0.5$ results and slightly higher at 300 Nmm$^{-3/2}$ for the $R = 0.1$ data (see Fig. 8). This division of the data is no longer apparent at 288°C.

For stress ratios up to 0.5, an increasing crack growth rate is observed when the temperature increases from 24 to 288°C, especially for low ΔK. For higher stress ratios, all the data fall within one scatter band.

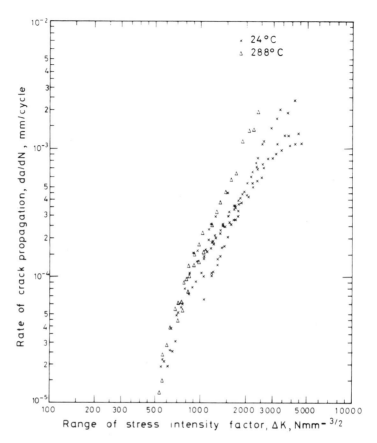

FIG. 7—*The effect of temperature for 1-in. thick specimens [27,35-36].*

Scatter in the Paris Law

As just seen, scatter may confuse the results obtained from a fatigue test so that trends in the data are difficult to establish. In probabilistic analyses, this scatter may be modeled by inserting distributions for C and m in the Paris law (Eq 1).

The problem appears to be slightly easier, since it has been found empirically that C and m are related—therefore only one distribution is required. Gurney [5] has analyzed several published crack propagation results for parent materials, weld metal, and heat affected zone materials. Considering the results for structural and high-strength steels together, the best fit relationship was found to be

$$C = \frac{1.315 \cdot 10^{-4}}{(895.4)^m} \tag{5}$$

This is for tests with the stress ratio approximately equal to zero.

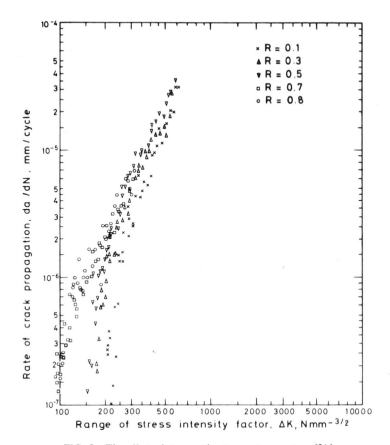

FIG. 8—*The effect of stress ratio at room temperature [21].*

This relationship may be used to derive a distribution for C as follows.

Since there are rarely sufficient data at one particular set of conditions to enable a statistical distribution to be found, an alternative approach is to pool data from varying tests. To do this, it is assumed that the distribution of C is the same for any particular value of m. The data are then transformed to one value, say $m = 3$. This is achieved by drawing an imaginary line through each pair (m', C') and parallel to the "best fit" line represented by Eq 5. Where this imaginary line crosses $m = 3$ gives one value of C in the required distribution.

Analyzing Gurney's data in this way, shows that a lognormal distribution provides a reasonable fit to the data. The appropriate parameters, for $m = 3$, are

$$\mu = -29.31 \quad \text{and} \quad \sigma = 0.24$$

(the mean and standard deviation of $\ln C$).

Clearly this is only a first approximation to the problem, but may well provide a valuable method as more data become available.

Example

In order to illustrate the calculation of a reliability figure, some of the preceding data have been used as an input to a probability analysis. In some places, simplifying assumptions have been made, which may not be the most realistic available. This is because the aim of this example is to produce a sample calculation rather than to expound some complicated numerical analysis.

The plain plate data at $-100°C$, reported earlier have been used to provide the toughness distribution, namely

$$\ell n\, \delta \cap N(-2.517, 0.504^2) \qquad (6)$$

that is, CTOD, δ is distributed lognormally with parameters μ and σ^2 equal to -2.517 and 0.504^2, respectively. Through fracture mechanics, this can be used to give a critical crack size distribution (a_c) [17].

In order to establish the failure probability, p_f, the distribution of a_c must be compared with that of the defect depth a_i, so that

$$p_f = \text{Prob}\,\{a_i > a_c\}$$

In service, initial defects may grow by fatigue, so that as cycles are accumulated, the overlap of the two distributions increases and hence p_f increases. (It is assumed here that the critical crack size distribution does not change with time.) If after N cycles have elapsed, the defect has grown to a depth $a(>a_i)$, then

$$p_f = \text{Prob}\,\{a > a_c\} > \text{Prob}\,\{a_i > a_c\}$$

To calculate p_f, a distribution for a must be derived—depending on that of a_i and the number of applied cycles, N.

Many fatigue cracks that occur in welded structures initiate from small, sharp slag intrusions situated at the weld toe. The average defect depth, a_i, is found to be 0.15 mm rising to a maximum of 0.4 mm. Since there is only a small chance of data falling more than three standard deviations above the mean value, it is assumed that a_i has mean 0.15 mm and standard deviation $\frac{1}{3}$ (0.4–0.15) = 0.083 mm. Translating these to the parameters of a lognormal distribution, which is usually used to model a_i gives

$$\ell n\, a_i \cap N(-2.031, 0.267) \qquad (7)$$

From the Paris Law for fatigue crack growth

$$\frac{da}{dN} = C(\Delta K)^m = C(Y \cdot \Delta\sigma\sqrt{\pi a})^m$$

where $\Delta\sigma$ is the applied stress range = 200 N/mm², in this case, and Y, a function of the crack and specimen geometry, is assumed here to be 1. Integrating

$$\int_{a_i}^{a} \frac{da}{a^{m/2}} = C(\Delta\sigma\sqrt{\pi})^m N$$

and

$$\ell n\, a = \ell n\, a_i + 1.257 \cdot 10^5 NC \qquad (8)$$

This is valid for $m = 2$ and shows that if $\ell n\, a_i$ and C are both normally distributed then so is $\ell n\, a$, where a is the crack length after N cycles. (Normality of C greatly simplifies the analysis.)

Fitting a normal distribution to Gurney's data, instead of the more realistic lognormal used earlier, gives

$$C \cap N(1.716.10^{-10}, 1.588.10^{-21}) \text{ for } m = 2 \qquad (9)$$

Combining Eqs 7, 8, and 9.

$$\ell n\, a \cap N(-2.031 + 2.157.10^{-5}N, 0.267 + 2.509.10^{-11}N^2) \qquad (10)$$

In terms of reliability calculations, the advantage of two lognormal distributions for crack length a and CTOD, δ has been discussed previously [17]. The resulting probability of failure, p_f may be calculated analytically with the aid of normal probability tables—no numerical computation is necessary.

$$p_f = \text{Prob}\{a > a_c\}$$

and this reduces [17] to $p_f = 1 - \Phi(\omega)$.

where

$$\omega = \frac{\ell n\, 122.586 - (\overline{\ell n\, a} - \overline{\ell n\, \delta})}{\sqrt{(\sigma_{\ell n\, a}^2 + \sigma_{\ell n\, \delta}^2)}}$$

and

$$\Phi(u) = \int_{-\infty}^{u} \frac{1}{\sqrt{2\pi}} \exp\{-\tfrac{1}{2}x^2\}\, dx$$

is tabulated for values of u.

The assumptions used to derive these expressions are detailed in the paper itself [*17*], with $\sigma_y = 430$ Nmm^{-2}, in this case.

Substituting for the parameters of the distributions of a and δ, from Eqs 6 and 10 gives

$$\omega = \frac{4.323 - 2.157 \cdot 10^{-5} N}{\sqrt{0.521 + 2.509 \cdot 10^{-11} N^2}}$$

Figure 9 represents the values of p_f for various values of N. From the form of ω, it can be seen that for small N up to 10^5, the initial defect distribution predominates, as expected. However, beyond 10^5, p_f increases to nearly 0.5 at $2 \cdot 10^5$ and hence to a 99.97 percent chance of failure after 10^6 cycles.

Concluding Remarks

The development of statistical reference fracture toughness curves has provided a means of predicting K_{Ic}-values from small-scale Charpy data. Applying this method to heats of material with varying chemical composition has indicated an improvement in toughness associated with a tighter control on the carbon, copper, and sulfur contents. Preliminary experimental data support this hypothesis.

Investigations into the statistical distributions for plain plate steel to BS4360 Grade 50D have indicated that lognormal or Weibull models may be used for crack tip opening displacement measurements at $-100°$C and at

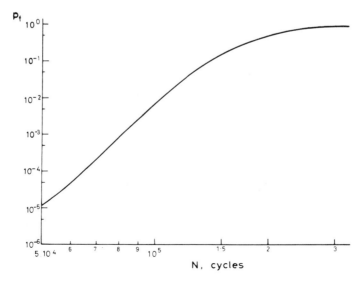

FIG. 9—*Variation of failure probability* p_f *with number of fatigue cycles, N.*

−60 and 0°C, respectively. Statistical analyses have been shown to yield estimates of tolerance limits and sample sizes associated with specified confidence levels.

The results of a qualitative study on fatigue crack propagation rates have been presented. The scatter in the data can be attributed in some cases to changes in the testing conditions, such as stress ratio, cycling frequency, and irradiation effects.

The distribution of C in the Paris Law has been found to be lognormal based on a variety of data for structural steels. This has been derived assuming a constant shape for all values of m and also that m and $\ln C$ are linearly related.

In conclusion, a simplified example demonstrates the calculation of a failure probability including a fatigue crack growth law, which radically alters the initial defect distribution after 10^5 cycles. A more realistic example could be analyzed using computer techniques, in particular Monte Carlo simulation.

Acknowledgments

This paper is published by permission of the Central Electricity Generating Board.

Part of this work was completed while at the Welding Institute. Their permission to include this work in the paper and the sponsorship by The Nuclear Installations Inspectorate of the analysis of fatigue crack propagation data are gratefully acknowledged.

References

[1] Oldfield, W. and Marston, T. U., *Proceedings*, 5th International Conference on SMiRT, Paper G2/7, Berlin, 1979.
[2] Oldfield, W., *Proceedings*, 6th International Conference on SMiRT, Paper G2/1, Paris, 1981.
[3] Johnston, G. O., "A Review of Probabilistic Fracture Mechanics Literature," to be published in *Reliability Engineering*, Sept. 1982.
[4] Paris, P. C. and Erdogan, F., *Transactions*, American Society of Mechanical Engineers, Vol. 85, Series D(4), 1963, pp. 528-534.
[5] Gurney, T. R., *Fatigue of Welded Structures*, Cambridge University Press, 1979, Chapter 2.
[6] Oldfield, W., Server, W. L., Wullaert, R. A., and Stahlkopf, K. E., *International Journal of Pressure Vessels and Piping*, Vol. 6, No. 3, 1978, pp. 203-222.
[7] Stahlkopf, K. E., Smith, R. E., Server, W. L., and Wullaert, R. A., *Transactions*, American Society of Mechanical Engineers, Vol. 99, No. 3, 1977, pp. 470-476.
[8] Server, W. L. and Oldfield, W., Nuclear Pressure Vessel Steel Data Base, EPRI Report NP—933, Project RP 886-1, 1978.
[9] Oldfield, W., Wullaert, R. A., Server, W. L., and Stahlkopf, K. E., *Proceedings*, 3rd International Conference on Pressure Vessel Technology, Tokyo, 1977, pp. 703-715.
[10] Marshall, W., "An Assessment of the Integrity of PWR Pressure Vessels," UKAEA Report by a Study Group under the Chairmanship of Dr. Marshall, 1976.
[11] Ingham, T. and Sumpter, J., *Proceedings*, Institute of Mechanical Engineers Conference on Tolerance of Flaws in Pressurised Components, London, 1978, p. 139.

[12] Ingham, T. and Morland, E., *Proceedings*, U.S. NRC, CSNI Specialists' Meeting on Plastic Tearing Instability, St. Louis, 1979, pp. 330-357.
[13] Krägeloh, E., Issler, L., and Zirn, R., *Proceedings*, 6th MPA Seminar, Stuttgart, Paper 8A, 1980.
[14] Druce, S. G. and Edwards, B. C., *Nuclear Energy*, Vol. 19, No. 5, 1980, pp. 347-360.
[15] Barsom, J. M. and Rolfe, S. T. in *Impact Testing of Metals, ASTM STP 466*, American Society for Testing and Materials, 1970, pp. 21-52.
[16] Johnson, N. L. and Leone, F. C., *Statistics and Experimental Design in Engineering and the Physical Sciences*, Wiley, New York, 1964.
[17] Johnston, G. O., *Proceedings*, 3rd National Reliability Conference, Birmingham, 1981, Paper 3C/1.
[18] Hald, A., *Statistical Theory with Engineering Applications*, Vol. 1, Wiley, London, 1952.
[19] Wilks, S. S., *Annals of Mathematical Statistics*, Vol. 13, 1942, pp. 400-409.
[20] Cullen, W. H. and Torronen, K., "A Review of Fatigue Crack Growth of Pressure Vessel and Piping Steels," NRL Memorandum Report 4298, NUREG/CR-1576.
[21] Bamford, W. H., Shaffer, D. H., and Jouris, G. M., *Proceedings*, 3rd International Conference on Pressure Vessel Technology, Tokyo, 1977, pp. 815-823.
[22] Paris, P. C., Bucci, R. J., et al in *Stress Analysis and Growth of Cracks, ASTM STP 513*, 1972, American Society for Testing and Materials, pp. 141-176.
[23] Stonesifer, F. R., *Engineering Fracture Mechanics*, Vol. 10, No. 2, 1978, pp. 305-314.
[24] Wilson, A. D., *Journal of Pressure Vessel Technology*, Vol. 99, No. 3, 1977, pp. 459-469.
[25] Fujii, E., Hayashi, S., and Iida, K., *Japan Welding Society*, Vol. 44, No. 8, 1975, pp. 13-21, translated by Addis Translations International, Calif.
[26] Legge, S. A. and Mager, T. R., HSSTP semi-annual progress report (period ending Aug. 1972), ORNL-4855, Oak Ridge National Laboratory, 1973, pp. 12, 15-16.
[27] Mager, T. R., *Proceedings*, 5th Annual HSST Information Meeting, Oak Ridge, 1971, Paper 1.
[28] Mager, T. R. and McLaughlin, V. J., HSST Technical Report 16, Westinghouse Electric Corporation, 1971.
[29] Mager, T. R., Landes, J. D., Moon, D. M., and McLaughlin, V. J., HSST Technical Report 35, Westinghouse Electric Corporation, 1973.
[30] Clark, W. G., Jr., *Journal of Materials*, Vol. 16, No. 1, 1971, pp. 134-149.
[31] Clark, W. G., Jr., and Wessel, E. T. in *Review of Developments in Plane Strain Fracture Toughness Testing, ASTM STP 463*, American Society for Testing and Materials, 1970, pp. 160-190.
[32] James, L. A. and Williams, J. A., *Journal of Nuclear Materials*, Vol. 47, 1973, pp. 17-22.
[33] James, L. A. and Williams, J. A., *Journal of Nuclear Materials*, Vol. 49, 1973/4, pp. 241-243.
[34] Kondo, T., Kikuyma, T., Nakajima, H., and Shindo, M., *Proceedings*, 6th Annual HSST Information Meeting, Oak Ridge, 1972, Paper 6.
[35] Mowbray, D. F., Andrews, W. R., and Brothers, A. J., *Transactions*, American Society of Mechanical Engineers, Journal of Engineering for Industry, 1968, pp. 648-655.
[36] Shahinian, P., Watson, H. E., and Hawthorne, J. R., *Journal of Engineering Materials and Technology*, Vol. 96, No. 4, 1974, pp. 241-248.
[37] Williams, J. A. and James, L. A., HSSTP semi-annual progress report (period ending Feb. 1973), ORNL-4918, Oak Ridge National Laboratory, 1973, pp. 31-41.

G. G. Trantina[1] and C. A. Johnson[1]

Probabilistic Defect Size Analysis Using Fatigue and Cyclic Crack Growth Rate Data

REFERENCE: Trantina, G. G. and Johnson, C. A., **"Probabilistic Defect Size Analysis Using Fatigue and Cyclic Crack Growth Rate Data,"** *Probabilistic Fracture Mechanics and Fatigue Methods: Applications for Structural Design and Maintenance, ASTM STP 798*, J. M. Bloom and J. C. Ekvall, Eds., American Society for Testing and Materials, 1983, pp. 67-78.

ABSTRACT: The fatigue failure mechanism of many cast metals and high-strength alloys involves initiation and propagation of a crack from an initial defect to a critical size where fatigue failure occurs. Crack growth rate and fatigue data can be combined to estimate the initial size of the defect that led to failure. This technique has been developed and used to provide statistical interpretations of fatigue data and to determine the probabilistic size distribution of critical defects and the specimen size dependence of the average critical defect size.

An expression for the fatigue curve that incorporates the initial defect size (a_i) is produced by integrating an analytical expression for the cyclic crack growth rate curve that includes a threshold stress intensity factor. For two different values of a_i, two fatigue curves are generated that can represent scatter in fatigue data from two specimens of the same size or average fatigue curves for specimens with two different sizes. This scatter or specimen size effect is best treated along lines of constant initial stress intensity factor that have a slope of -0.5 on the log stress-log life fatigue curve. The separation between fatigue curves along lines of constant initial stress intensity factor is independent of fatigue life. This treatment of fatigue data therefore predicts the commonly observed trend of increasing scatter in fatigue life with increasing lifetime and provides a treatment of scatter that is independent of fatigue life.

Probabilistic defect size analysis has been applied to data for a nickel-base superalloy to predict the size distribution of critical defects. An initial defect size is inferred for each data point, then a distribution of initial critical defect sizes is predicted for the alloy. The Weibull distribution is used to display the data in terms of $1/\sqrt{a_i}$. The specimen size effect is treated where the square root of the average critical defect size increases with an increase in the effective area of the specimen. Thus, critical defect size distributions in components can be predicted that, in turn, allow material processing and inspection criteria to be based on required component life.

---

[1]Mechanical engineer, and ceramist, respectively, General Electric Company, Corporate Research and Development, Schenectady, N.Y. 12301.

KEY WORDS: fracture mechanics, probabilistic fracture mechanics, fatigue (materials), cyclic crack growth, defect size, stress intensity factor, scatter, size effect, Weibull distribution, fatigue crack growth

Probabilistic techniques must be developed as a means of predicting the lifetimes of components. As new materials are developed and tested, new design techniques must be used to relate the mechanical behavior of materials to the performance of components. A statistical fatigue failure analysis based on the Weibull weakest link concept has been suggested by Trantina [1][2] to provide an interpretation of size effects in fatigue behavior. It is presumed that the variation or scatter in the fatigue life or fatigue strength (stress to cause failure for a particular lifetime) is dependent only on the variation of the size of the strength controlling defect. When a distribution of defect sizes is present in a material, the probability of sampling a large defect is small if one samples or tests only a small amount of material. The larger the amount of material sampled, the greater the probability that it contains a large defect. Thus, the size of the most severe defect will, on the average, increase and the fatigue strength will decrease with increasing volume or area under stress. A weakest link model was used to characterize the behavior where the critical flaw or the "weakest link" of the material provides the crack initiation source [1]. This approach of probabilistic fatigue analysis does not require information concerning the size distribution of critical defects or the crack growth rate behavior.

The fatigue failure mechanism of many materials involves initiation and propagation of a crack from an initial defect (pore, inclusion, etc.) to a critical size where fatigue failure is defined [2]. An analysis of this process involves three types of material data—cyclic crack growth rates, fatigue (S-N), and initial defect size. Any two of these three types of data can be used to predict the third [2,3]. In the analysis presented in this paper, a relationship is fit to crack growth rate data. For an initial defect size, the relationship is integrated to produce a fatigue curve. An initial defect size is thus inferred for a fatigue data point that falls on the curve. Therefore a distribution of critical initial defect sizes can be predicted from a set of fatigue data. Again, it is presumed that the variation or scatter in fatigue data is dependent only on the variation of the initial defect size and not on metallurgical inhomogeneity or other sources. For some materials with a small amount of scatter in fatigue data, sources other than initial defect size become important. However, for many cast metals and high-strength alloys where the fatigue failure mechanism is propagation of a crack from an initial defect [3], significant scatter arises primarily from variation of the initial defect size.

The objectives of this paper are to describe the approach and demonstrate its application to a nickel-base superalloy by determining both the size distri-

[2]The italic numbers in brackets refer to the list of references appended to this paper.

bution of critical initial defects and the specimen size effect on the average defect size. By relating fatigue life to critical defect sizes in test specimens, critical defect size distributions in components can be predicted, thus allowing material processing and inspection criteria to be based on required component life.

Cyclic Crack Growth Analysis

The crack growth per cycle (da/dN) is controlled by the range of stress intensity factor (ΔK) during cyclic loading. The stress intensity factor is a measure of the magnitude of the stress field at the crack tip and is computed from the applied stress range ($\Delta \sigma$) and the crack length (a)

$$\Delta K = Y \Delta \sigma \sqrt{a} \tag{1}$$

where Y is a geometric factor that accounts for the geometry of crack and structure. The crack growth rate can be expressed as a power law

$$\frac{da}{dN} = A(\Delta K - \Delta K_{th})^n \tag{2}$$

where A, n, and ΔK_{th} are constants. The threshold stress intensity range, ΔK_{th}, is added as the third constant to account for the variable slope of a typical crack growth rate curve (Fig. 1). For many materials, da/dN becomes very small at small ΔK's, indicating an apparent threshold stress intensity factor

$$\Delta K_{th} = Y \Delta \sigma \sqrt{a_{th}} \tag{3}$$

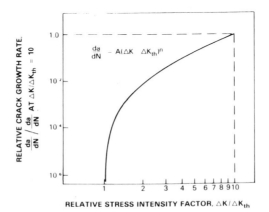

FIG. 1—*Cyclic crack growth rate curve.*

where a_{th} is the threshold crack length for an applied stress range ($\Delta\sigma$). For $a > a_{th}$, cracks will propagate, and for $a < a_{th}$, cracks will not propagate.

The crack growth rate expression (Eq 2) can be used to predict the number of cycles-to-failure (N_f) assuming that:

1. The fatigue failure process consists entirely of the propagation of a crack from an initial size to a critical size.
2. The shape of the crack remains constant during propagation of the crack.
3. The crack driving force is characterized by the elastic stress intensity factor.
4. The stress range, $\Delta\sigma$, is constant during the crack propagation.
5. The mean stress remains constant during crack propagation.

The crack growth rate expression can be integrated by substituting Eqs 1 and 3 into Eq 2 and separating the variables

$$\int_{a_i}^{a_f} \frac{da}{(\sqrt{a} - \sqrt{a_{th}})^n} = A Y^n \Delta\sigma^n \int_0^{N_f} dN \qquad (4)$$

The initial crack length (a_i) is defined in terms of the initial stress intensity factor

$$\Delta K_i = Y \Delta\sigma \sqrt{a_i} \qquad (5)$$

and the final crack length (a_f) is defined by the critical value of K

$$K_c = Y \Delta\sigma \sqrt{a_f} \qquad (6)$$

The assumed shape of the crack is a semicircular surface flaw of length a where $Y = 1.12 \, (2/\pi) \, (\sqrt{\pi})$. The cyclic loading is from zero to the maximum load. The effect of other stress intensity ranges ($R = K_{min}/K_{max}$) can be included [3]. Integrating Eq 4 yields the number of cycles-to-failure

$$N_f = \frac{2}{A Y^n \Delta\sigma^n} \left[\frac{(\sqrt{a_f} - \sqrt{a_{th}})^{2-n}}{2 - n} + \sqrt{a_{th}} \frac{(\sqrt{a_f} - \sqrt{a_{th}})^{1-n}}{1 - n} \right.$$

$$\left. - \frac{(\sqrt{a_i} - \sqrt{a_{th}})^{2-n}}{2 - n} - \sqrt{a_{th}} \frac{(\sqrt{a_i} - \sqrt{a_{th}})^{1-n}}{1 - n} \right] \qquad (7)$$

The fatigue curve is thus a function of the three constants from the crack growth rate expression (A, n, ΔK_{th}), the fracture toughness (K_c), the crack geometry (Y), and the initial crack length (a_i). Of these parameters, the initial crack length is probably the most difficult to determine and has the most sensitive effect on the fatigue life. Another form of Eq 7 that is useful in understanding the parameters that control fatigue life is

$$N_f = \frac{2}{A Y^2 \Delta\sigma^2} \left[\frac{(K_c - \Delta K_{th})^{2-n}}{(2 - n)} + \Delta K_{th} \frac{(K_c - \Delta K_{th})^{1-n}}{(1 - n)} \right.$$
$$\left. - \frac{(\Delta K_i - \Delta K_{th})^{2-n}}{(2 - n)} - \Delta K_{th} \frac{(\Delta K_i - \Delta K_{th})^{1-n}}{(1 - n)} \right] \quad (8)$$

In this expression, it can be seen how the initial stress intensity factor (ΔK_i) affects the fatigue life. The number of cycles to failure is a function of several constants but only two variables, $\Delta\sigma$ and ΔK_i. In the following section, the process of inferring a size distribution of critical defects by using Eq 7 and the implications of Eq 8 on scatter in fatigue data will be discussed.

Analysis of Fatigue Data

The scatter in fatigue data is assumed to arise only from a variation in initial defect size and not from metallurgical inhomogeneity or other sources. A fatigue curve can be produced by using Eqs 7 or 8 and by assuming a set of material properties (A, n, ΔK_{th}, K_c), a crack geometry (Y), and an initial crack size (a_i). In Fig. 2, two such curves are shown for two values of a_i. These two curves could represent scatter in fatigue data from two specimens of the same size but with different initial defect sizes, or they could represent average fatigue curves for specimens with two different sizes where the larger specimens would have, on the average, a larger critical defect size and a shorter fatigue life. Treating the

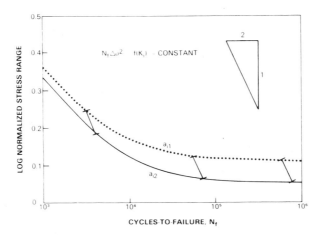

FIG. 2—*Fatigue curve for two critical initial defect sizes.*

scatter or specimen size effect as variability in fatigue life at constant stress results in increasing variability with increasing fatigue life (Fig. 2). At long lifetimes, the scatter becomes enormous and life prediction is complicated by tests that have been terminated without failure. This increased scatter with increased lifetime is well-known and has been quantified for low- and high-strength steel, aluminum, and titanium [4]. A more satisfactory method treats scatter and specimen size effects as variability in fatigue strength at constant fatigue life [1]. This results in only a small increase in the scatter or specimen size effect of the fatigue strength with increasing fatigue life (Fig. 2). Nevertheless, treatment of scatter either at constant life or at constant stress results in variability or scatter that is different from point to point on Fig. 2. The following describes a treatment of fatigue data where the variability is constant throughout the S-N curve.

Equation 8 can be expressed as

$$N_f \Delta\sigma^2 = f_1(\Delta K_i) = f_2(\Delta\sigma\sqrt{a_i}) \tag{9}$$

From this, one can see that lines of constant ΔK_i have a slope of -0.5 on plots of log $\Delta\sigma$ versus log N_f such as Fig. 2. A constant value of ΔK_i can result from a small crack and a large stress or a large crack and a small stress (Eq 9). In either case, the initial response to the crack driving force (ΔK) is the same since the starting point (ΔK_i) on the crack growth rate curve is the same. The lifetimes are different because $N_f \Delta\sigma^2 = $ constant (Eq 9).

It can be shown that fatigue curves such as the two on Fig. 2 are identical in shape but offset along lines that also have a slope of -0.5. The distance separating the two curves measured along lines with a slope of -0.5 correspond to the length of the hypotenuse of the triangle shown. This length, L, for an arbitrary slope is

$$L^2 = [\log(N_2/N_1)]^2 + [\log(\Delta\sigma_1/\Delta\sigma_2)]^2 \tag{10}$$

The length along a slope of -0.5 can be expressed in terms of $\Delta\sigma$'s only

$$L = \sqrt{5}\log\left(\frac{\Delta\sigma_1}{\Delta\sigma_2}\right) \tag{11}$$

As just described, lines of slope $= -0.5$ are lines of constant ΔK_i, therefore, combining Eqs 5 and 11 at constant ΔK_i yields

$$L = \sqrt{5/2}\log(a_{i2}/a_{i1}) \tag{12}$$

This distance, L, is only a function of the ratio of initial flaw sizes and is independent of $\Delta\sigma$, N_f, A, n, ΔK_{th}, ΔK_c, and Y. The distance between the two curves is constant and independent of position, therefore, the two curves are identical in shape and displaced along lines with a slope of -0.5.

The distribution of critical flaw sizes in a material can be pictured as a family of curves such as the two on Fig. 2. The relative positions of the curves are identical along all lines with a slope of -0.5; therefore, the scatter or specimen size effect is independent of stress and life when analyzed along such lines. In summary, scatter and size effects should be treated along lines of constant initial stress intensity factor (ΔK_i).

To infer a distribution of critical initial defect sizes from fatigue data and cyclic crack growth rate data, each fatigue data point ($\Delta \sigma$, N_f) must be associated with a fatigue curve. The fatigue curves are produced by Eq 7 with appropriate material properties to describe the crack growth rate. By iteratively varying a_i, a fatigue curve can be found that passes through a fatigue data point. The associated a_i for that curve is then an estimate of the initial size of the defect that led to failure in that specimen. This technique is illustrated schematically in Fig. 3 for three fatigue data points.

Distribution of Defect Sizes

After the initial sizes of the strength-controlling defects have been computed for a data set, a mathematical representation must be used to statistically treat the size distribution. The two-parameter Weibull distribution function [5] has been chosen

$$P = 1 - \exp\left[-\int\left(\frac{\Delta\sigma}{\Delta\sigma_0}\right)^m dA\right] \quad (13)$$

where
- P = cumulative probability of failure,
- m = Weibull modulus, and
- $\Delta\sigma_0$ = a normalizing constant.

The integration of $(\Delta\sigma/\Delta\sigma_0)^m$ is carried out over all elements of surface (dA) if the defects leading to fatigue failure are located only on the surface. Other-

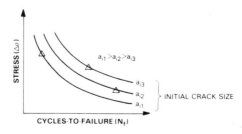

FIG. 3—*Three fatigue curves passing through three data points.*

wise, the integration is carried out over the specimen volume. The Weibull modulus is a measure of the scatter or variability in defect sizes where a small value of m indicates a large degree of variability. Accordingly, the Weibull modulus describes the variability or scatter in fatigue strength and the sensitivity of the fatigue strength to specimen size. Equation 13 can be expressed as

$$P = 1 - \exp\left[-kA\left(\frac{\Delta\sigma_{max}}{\Delta\sigma_0}\right)^m\right] \quad (14)$$

where
$\Delta\sigma_{max}$ = maximum $\Delta\sigma$ on the surface, and
k = a dimensionless constant defined as

$$k = \int^A \left(\frac{\Delta\sigma}{\Delta\sigma_{max}}\right)^m \frac{dA}{A} \quad (15)$$

Tension fatigue specimens with no stress concentrations ($K_t = 1$) and uniform stress throughout result in $k = 1$. All other stress states result in $0 < k < 1$. The product of k times A in Eq 14 is often described as the "effective area" and is the area of the specimen that is effectively being fatigued at $\Delta\sigma_{max}$.

Since scatter and size effects should be treated at constant ΔK_i (and therefore constant $\Delta\sigma\sqrt{a_i}$), $\Delta\sigma$ in Eq 14 can be replaced by $1/\sqrt{a_i}$. A graphical method of displaying the distribution as a straight line relationship involves a rearrangement of Eq 14 to

$$\ell n \, \ell n \, \frac{1}{1-P} = m \, \ell n \, \frac{1}{\sqrt{a_i}} + \text{constant} \quad (16)$$

If one plots

$$\ell n \, \ell n \, \frac{1}{1-P} \text{ versus } \ell n \, \frac{1}{\sqrt{a_i}},$$

the data points should fall along a straight line with a slope equal to the Weibull modulus, m. The probability of failure, P, for each specimen can be estimated as $(n - 0.5)/N$, where n is the ordering number (weakest to strongest or largest to smallest crack length) and N is the total number of specimens.

An example of a size distribution of critical surface defects is shown in Fig. 4 for 20 smooth (unnotched) specimens of a nickel-base superalloy. The cumulative probability of failure is displayed versus $1/\sqrt{a_i}$ where each a_i is normalized by the average value of a_i for the data set. The Weibull distribution provides a reasonable fit of the size distribution of critical defects. The slope of the linear regression line through the data is 11. The data indicates that

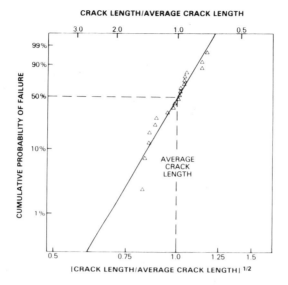

FIG. 4—*Size distribution of critical initial defects.*

the weakest of 100 specimens has an initial defect length equal to about twice the average.

Specimen Size Effect

The specimen size effect in fatigue occurs because the size of the critical defect, on the average, increases with increasing area or volume or with more uniform stress distributions (larger k). This size effect also can occur between specimens and components. From Eq (14), one can show that at constant probability of failure

$$\frac{\Delta\sigma_1}{\Delta\sigma_2} = \left(\frac{k_2 A_2}{k_1 A_1}\right)^{1/m} \qquad (17)$$

Thus, for $\Delta K_i =$ constant

$$\frac{\sqrt{a_{i2}}}{\sqrt{a_{i1}}} = \left(\frac{k_2 A_2}{k_1 A_1}\right)^{1/m} \qquad (18)$$

The specimen size effect is now given in terms of the square root of the initial average defect size.

The size effect for a nickel-base superalloy is demonstrated by comparing the smooth specimen results in Fig. 4 with results from notched bars. All of the failures were from surface flaws. The effective area of notched bars is smaller than smooth bars because of the nonuniform stress distribution. The effective area decreases as the stress concentration factor (K_t) increases, since the stress distribution along the face of the notch becomes more nonuniform. The size distributions of critical defects for the smooth bars and three sets of notched bars with different K_t's are shown in Fig. 5. The scatter in the four size distributions of critical defects is nearly the same (m's of 11, 14, 13, and 18 for K_t's of 1.0, 1.6, 1.8, and 2.5, respectively) with an average m of about 14. The average size of the defects for the four specimen types can be displayed as a function of the effective areas for the four specimen types as shown in Fig. 6. The m-value from the slope of the line in Fig. 6 (Eq 18) is about 18, which agrees reasonably well with the preceding estimates. The scatter about the average behavior is also illustrated on Fig. 6 as the weakest of 100 specimens. Information such as that shown on Figs. 5 and 6 can be used to predict the size distribution of critical defects for other specimen types or for structures.

Conclusions

For an initial critical defect size (a_i), a fatigue curve can be produced by integrating the analytical expression for a cyclic crack growth rate curve. To per-

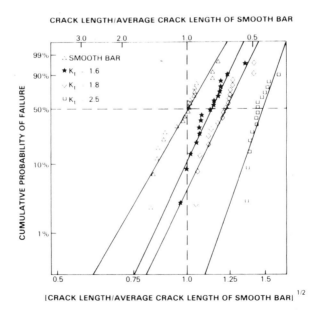

FIG. 5—*Size distribution of critical initial defects in smooth and notched bars.*

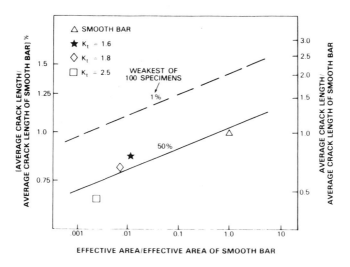

FIG. 6—*Specimen size effect of average critical defect size.*

form this calculation, a number of assumptions were made. First of all, the crack propagation assumption is justified for most cast metals and high-strength alloys since the fatigue failure mechanism is propagation of a crack from an initial defect. This initial defect size has the most sensitive effect on fatigue life when compared to the crack growth rate constants, the fracture toughness, and crack shape. For the data considered in this analysis, the crack driving force can essentially be characterized by the elastic stress intensity factor, and the stress range and mean stress remain relatively constant during crack propagation.

For two different values of a_i, two fatigue curves can be generated to represent scatter in fatigue data from two specimens of the same size or average fatigue curves for specimens with two different sizes. This scatter or specimen size effect should be treated along lines of constant initial stress intensity factor. These lines have a slope of -0.5 on the log stress-log life fatigue curve. Along these lines the scatter or size effect in stress or life is independent of fatigue life. An initial defect size can be inferred for a data point that falls on the fatigue curve. Then, a size distribution of critical defects can be predicted for a set of fatigue data. The Weibull distribution can be used to display the data in terms of $1/\sqrt{a_i}$. The specimen size effect can be treated where the square root of the average size of the distribution of critical defects increases with effective area. Thus, critical defect size distributions in components can be characterized so that material processing and inspection criteria can be based on required component life.

References

[1] Trantina, G. G., *Journal of Testing and Evaluation*, Vol. 9, No. 1, Jan. 1981, pp. 44-49.
[2] Paris, P. C., "The Fracture Mechanics Approach to Fatigue," *Proceedings*, 10th Sagamore Conference, Syracuse University Press, Syracuse, 1964.
[3] Kaisand, L. R. and Mowbray, D. F., *Journal of Testing and Evaluation*, Vol. 7, No. 5, Sept. 1979, pp. 270-280.
[4] Freudenthal, A. M., *Journal of Aircraft*, Vol. 14, No. 2, Feb. 1977, pp. 202-208.
[5] Weibull, W., "A Statistical Theory of the Strength of Materials," *Proceedings*, Royal Swedish Institute of Engineering Research, No. 151, 1939.

A. P. Berens[1] *and P. W. Hovey*[1]

Statistical Methods for Estimating Crack Detection Probabilities

REFERENCE: Berens, A. P. and Hovey, P. W., **"Statistical Methods for Estimating Crack Detection Probabilities,"** *Probabilistic Fracture Mechanics and Fatigue Methods: Applications for Structural Design and Maintenance, ASTM STP 798*, J. M. Bloom and J. C. Ekvall, Eds., American Society for Testing and Materials, 1983, pp. 79-94.

ABSTRACT: To characterize the uncertainty in nondestructive evaluation (NDE), a probability of crack detection (POD) as a function of crack length is postulated where POD is defined as the proportion of cracks of a given length that would be detected by the NDE technique when applied by inspectors to structural elements in a defined environment. This paper presents a statistical framework for describing the uncertainty in the NDE determinations, and evaluates various characterizations of NDE reliability. The data from a recent Air Force study on NDE reliability are used to estimate the parameters of the NDE model. For these representative capabilities, NDE reliability experiments are simulated. Different NDE capability characterizations are computed for each simulated experiment and are statistically compared.

KEY WORDS: nondestructive evaluation, NDE reliability, NDE capability, probability of crack detection, crack detection reliability, inspection reliability, probabilistic fracture mechanics, fatigue (materials), fatigue crack growth, fracture mechanics

The United States Air Force approach to preventing structural fatigue failures is based on predictions of potential crack length as a function of flight time [1-4].[2] For each critical area of the structure: (1) the length of a potential crack that could be present is ascertained; (2) the growth of this crack is predicted for the anticipated usage environment; (3) a calculation of structural strength degradation due to crack growth is utilized to determine the safe time period during which the structure will not fail; and (4) a repeat inspection is scheduled at half the flight time required for the crack to grow to critical size. The assumed crack length in new structure is representative of manufacturing quality while the reset crack length at an inspection, a_{NDE}, is the longest crack

---

[1] Senior research statistician and associate research statistician, respectively, Service Life Management, Aerospace Mechanics, University of Dayton Research Institute, Dayton, Ohio 45469.
[2] The italic numbers in brackets refer to the list of references appended to this paper.

that could pass undetected through the nondestructive evaluation (NDE) system. Since crack growth rates are highly dependent on crack length, the success of this procedure is greatly influenced by the correct choice of the initial and reset crack lengths.

A characteristic of all current nondestructive evaluation techniques is their inability to repeatedly produce correct indications when applied by various inspectors to flaws of the same "size" [5-7]. The ability and attitude of the operator, the geometry and material of the structure, the environment in which the inspection takes place, and the location, orientation, and size of the flaw all influence the chances of detection. However, since the structural maintenance actions are scheduled on the basis of potential crack length, the other factors are controlled (to the extent possible) or ignored, and the resulting inspection uncertainty is characterized only in terms of crack length. A probability of crack detection (POD) for all cracks of a given length is postulated as the proportion of cracks that will be detected by an NDE system when applied by representative inspectors to a population of structural elements in a defined environment. Thus, the capability of an NDE system is expressed in probabilistic terms and this characterization has two significant ramifications.

First, for a given NDE application, the true probability of detection as a function of crack length (or for a single crack length) will never be known exactly. The capability of an NDE system can only be demonstrated through an experiment in which representative structures with known crack lengths are inspected and the true POD is estimated by the observed percentage of correct positive indications. The estimated POD is subject to statistical variation that results from finite sample sizes in the experiment and, possibly, systematic differences between experimental and in-service conditions. However, statistical methods (which depend on the experimental procedure) are available that yield confidence limits on the true probability to account for finite sample size. Thus, some protection against making a wrong decision on the basis of a set of nontypical results is provided by the confidence limits but an unknown element of risk due to systematic experimental errors may be present.

Second, in the real world structural integrity problem, no inspection procedure will provide 100 percent assurance that all cracks greater than some useful length will be detected. The current capabilities and the uncertainty resulting from the NDE demonstration process at the short crack lengths of interest in aircraft applications dictate that the a_{NDE} value must be specified in terms of a high confidence (CL) that a high percentage (POD) of all cracks greater than a_{NDE} (the POD/CL limit) will be found. For example, MIL-A-83444 [2] indicates that a_{NDE} is that crack length for which it can be shown there is 95 percent confidence that 90 percent of all cracks will be found (that is, the 90/95 crack length). Note that the chances of a crack longer than a a_{NDE} passing undetected depends not only on the capability of the NDE system but also on the distribution of crack lengths that are present in the structural elements before the inspection.

Due to the importance of the statistical characterization of NDE capabilities, this study was undertaken to: (1) evaluate and compare existing methods for determining the POD as a function of crack size; (2) to devise and evaluate different analysis methods and models; and (3) to evaluate different combinations of POD and confidence limits as single number characterizations (a_{NDE} values) of NDE systems. The essential tools used to achieve these objectives were the formulation of a probabilistic framework that recognizes that different cracks of the same length have different probabilities of detection and the use of this formulation to simulate NDE reliability demonstration experiments.

It should be noted that in general, NDE reliability comprises two types of wrong indications: failure to give a positive indication in the presence of a crack and giving a positive indication when there is not a crack (a false call). While it is recognized that both types of error are important, only the former is considered in this study. Due to the potential safety hazard, failure to find a crack is considered to be the more critical problem in the structural integrity application.

NDE Reliability Experiments

There are three categories of experiments that have been used to evaluate the reliability of NDE systems: (1) demonstration of a capability at one crack length; (2) estimation of the POD function and confidence bounds through single inspections of cracks covering a range of crack lengths; and (3) estimation of the POD function and confidence bounds through multiple inspections of cracks covering a range of lengths. Analysis of data from Categories 1 and 2 experiments have generally been based only on binomial distribution theory [5,6]. Category 3 data have been analyzed by fitting an equation to the observed detection probabilities of the cracks [7]. The following paragraphs briefly summarize the binomial analysis method and demonstrate that a regression model is appropriate for characterizing the POD function.

Binomial Method

In the binomial method of analysis, the cracks in the experimental specimens are grouped into intervals as defined by crack length. Assume there are n_i cracks in interval (a_{i-1}, a_i) and that r_i of the cracks are discovered during inspection. If it is assumed that all cracks in the interval have a probability, p_i, of being detected, then p_i can be estimated by

$$\hat{p}_i = r_i/n_i \tag{1}$$

and binomial distribution theory can be used to calculate confidence limits for the true, but unknown, value of p_i. The confidence limits depend on both r_i

and n_i and are tabled in many references [5,8]. The lower confidence bound for the POD is usually assigned to the crack length at the upper end of each interval, a_i, to provide an added measure of conservation.

Since the lower confidence bound is strongly dependent on the number of cracks in an interval, different schemes have been devised for assigning cracks to intervals so that the lower confidence bound is as well behaved as possible function of crack length. Yee et al [6] devised an algorithm for grouping results to achieve the highest possible lower confidence bound on the estimates of p_i. This method, called the optimized probability method (OPM) was recommended over more simple grouping schemes. Thus, the OPM was selected in this paper as the representative binomial method for comparison with the regression approach.

Regression Method

In a large NDE reliability program conducted for the Air Force, sections of retired aircraft and other specimens were transported to Air Force depots and inspected for cracks by representative personnel, using various NDE systems in a typical inspection environment [7]. At the completion of the travel phase of the program, the structures were inspected destructively to verify the existence and lengths of the cracks. This experiment has often been called the "Have-Cracks-Will-Travel" Program and the data is called the "Have Cracks" data.

Since there are multiple inspections for each crack, this method of collecting data yields an estimate of a detection probability for each individual crack. In the Have Cracks data, there are many examples of two cracks of approximately equal length having significantly different detection probabilities (see, for example, Fig. 2). These results clearly demonstrated that the POD is influenced by many factors and the correlation with crack length is not necessarily strong. To model the POD, it can be assumed that there is a distribution of detection probabilities at each crack length where the scatter in this distribution is caused by the nonreproducibility of all factors other than crack length. Examples of such factors are differences in detectability due to operators, human factors, environments, and crack orientation, geometry or location.

Let $f_a(p)$ represent the density function of the detection probabilities for the population of details that have a crack length of a. Figure 1 presents a schematic representation of this distribution. The probability of selecting a crack whose detection probability is between p and $p + dp$ is $f_a(p)dp$ by definition of a density function. The probability that this crack will be detected is p. Thus, the conditional probability that the detection probability is p and that a positive indication will be given is $pf_a(p)dp$. To find the unconditional probability that a randomly selected crack of length a will be detected, POD(a), the conditional probabilities are summed over the range of detection probabilities. Therefore

$$\text{POD}(a) = \int_0^1 p f_a(p) \, dp \qquad (2)$$

Since Eq 2 is also the formula for the mean of the distribution of detection probabilities at crack length a, POD (a) is also the traditionally defined regression of detection probabilities on crack length. Therefore, regression analysis techniques can be used to estimate and place confidence limits on POD(a) when individual estimates of the detection probabilities are available.

Although POD (a) is the curve through the means of the detection probabilities, a functional model for this curve must be determined or assumed before the characterization can be used. Since the Have Cracks data are representative of field inspection capability for selected structures and inspection methods and are the largest data set for which detection probabilities have been estimated for many cracks from many inspectors, they were selected for a study to determine an acceptable model for the POD (a) function. Three criteria were established for the definition of "acceptable": (1) goodness of fit, (2) normality of deviations from fit, and (3) equality of variance of deviations from fit for all crack lengths. The latter two criteria are necessary statistical properties for the validity of confidence limits derived from regression analyses.

Seven functional forms were evaluated with respect to the criteria for acceptability. These included the Weibull, modified Weibull [2], probit, log probit, logistics, log logistics, and the arcsine models. The last five were selected due to their widespread use in an analogous problem of modeling percent response as a function of dosage in the field of bioassay [9,10]. All of the models were applied by performing appropriate transformations to the detection probabilities, p_i, and the crack lengths, a_i, and fitting the transformed variables with a linear regression analysis. The deviations of the transformed observations

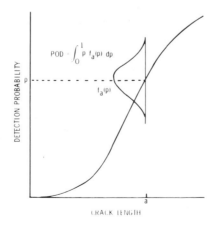

FIG. 1—*Schematic representation of distribution of detection probabilities for cracks of fixed length.*

from the regression equations were analyzed to test the applicability of each model [11].

The analysis of the models indicated that the log logistics (hereafter called the log odds) model was best among those tested. Moreover, the log odds model was generally acceptable for the Have Cracks data sets. The goodness of fit was always acceptable and equality of variance was never rejected but normality of deviations from the model was rejected for two of the seven data sets at a level of significance between 0.05 and 0.10. Therefore, this model was used in the subsequent evaluation of analysis methods.

The functional form of the log odds model is given by

$$\text{POD}(a) = \frac{\exp(\alpha + \beta \ln a)}{1 + \exp(\alpha + \beta \ln a)} \quad (3)$$

or

$$\ln\left[\frac{\text{POD}(a)}{1 - \text{POD}(a)}\right] = \alpha + \beta \ln a \quad (4)$$

Given the pairs of data points $(a_i, \hat{p}_i)$, let x_i and y_i be defined by

$$x_i = \ln a_i \quad (5)$$

$$y_i = \ln\left[\frac{\hat{p}_i}{1 - \hat{p}_i}\right] \quad (6)$$

α and β can be estimated as the well-known linear regression estimates of y on x and confidence bounds can be placed on the mean y for fixed x [12]. Since the log odds transformation is monotonic, the reverse transformation of the confidence bound is the confidence bound on POD(a). Figure 2 is an example application for inspections of cracks emanating from fastener holes in a skin and stringer wing assembly using eddy current surface scans [7,11]. Each data point represents one crack with its associated length and observed detection probability in 60 independent inspections. The solid curve is the estimate of the log odds POD(a) function and the dashed curve is the lower 95 percent confidence bound on POD(a). The minimum crack length for which there is a 95 percent confidence that 90 percent of the cracks will be detected (the 90/95 limit) is the crack length for which the confidence limit curve equals 90 percent. For the data of Fig. 2, the 90/95 limit is approximately 25 mm.

Evaluation of NDE Analysis Methods

To evaluate and compare the various methods for analyzing data from NDE capability demonstration programs, results from a large number of experi-

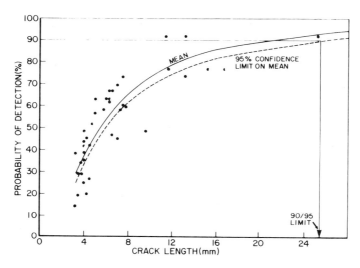

FIG. 2—*Regression analysis—eddy current inspections of skin and stringer wing assembly, 60 inspections per fastener hole.*

ments under known conditions are necessary. Since such tests are expensive and the experimental conditions are difficult to hold fixed over the long intervals necessary to repeat experiments, "experimental" NDE data were generated in a Monte Carlo computer simulation of inspections. The simulation process enabled the generation of a large number of NDE experiments under a "known capability" and under selected changes in experimental conditions.

Simulation Procedure

Figure 3 presents a schematic diagram for the process of simulating NDE experiments. The process simulates one NDE experiment by simulating the results of one inspection of each of 400 details with cracks of different lengths. The simulation of the inspection of each detail requires three steps.

1. To reflect the random nature of the crack sizes that may be present in the details to be inspected, a distribution of crack sizes is assumed. A simulated inspection is initiated by selecting a crack size at random, a_i, from this distribution. The assumed crack size distribution was considered to be an experimental condition.

2. A detection probability is determined for the crack using the model

$$Y_i = \alpha + \beta \ln a_i + \epsilon_i \tag{7}$$

where ϵ_i is randomly selected from a normal distribution with zero mean and standard deviation, $S(\epsilon)$, chosen to reflect the variability of the detection prob-

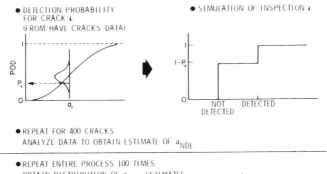

FIG. 3—*Schematic of procedure for simulating NDE experiments.*

abilities about the POD curve. To simulate the Have Cracks data, $S(\epsilon)$ would be the standard error as determined during the regression analysis. Having randomly determined ϵ_i, the detection probability for the crack is given by

$$p_i = \frac{\exp(\alpha + \beta \ln a_i + \epsilon_i)}{1 + \exp(\alpha + \beta \ln a_i + \epsilon_i)} \tag{8}$$

3. Given the detection probability, p_i, for the crack, a simple Bernoulli trial is simulated with probability of p_i of successfully detecting the crack and $(1 - p_i)$ of failing to detect it. The result of the "inspection" is recorded either as $(a_i, 1)$ if the "crack" was "detected" or $(a_i, 0)$ if the "crack" was not "detected."

After the preceding steps are repeated 400 times to complete an entire experiment, the data were analyzed by the different analysis methods. These included the methods based on the binomial distribution and the regression analysis using the log odds model. Upper confidence limits on crack sizes (POD/CL limits) were calculated using all combinations of POD equal to 0.5, 0.75, 0.9, 0.95, and 0.99 and confidence limits for 90, 95, and 99 percent.

The preceding procedure was repeated to generate 100 repetitions of the basic NDE reliability experiment and the associated estimates of the POD(a) function and the POD/CL values. These data formed the basis for the comparison and evaluation of the various methods for estimating the capability of an NDE system.

While all of the large data sets of the Have Cracks data have been simulated [11], this paper focuses only on the simulation of the eddy current surface scans around countersunk fasteners in a skin and stringer wing assembly. This data set is typical of the Have Cracks data. The original data points are shown in Fig. 2. The mean (solid) line of Fig. 2 was taken as the true POD(a) and for this curve $\alpha = -2.9$ and $\beta = 1.7$. For this NDE capability, 90 percent of 20.1 mm cracks and 95 percent of 31.1 mm cracks will be detected.

Two crack size distributions were used in the simulated experiments. First, the distribution of the reported crack sizes of the original Have Cracks data set were employed but these crack sizes were too short for use in the binomial analysis method for the NDE capability assumed. These cracks had a median of 5.8 mm and a 90th percentile of 12.7 mm. To effect a comparison between the binomial and regression analysis methods, a distribution of longer cracks was introduced. This distribution was arbitrarily assumed to be lognormal with a median crack length of 12.7 mm and 90th percentile of 76.2 mm.

Most of the simulations were performed using the scatter in detection probabilities about the POD(a) as displayed in Fig. 2. However, to evaluate the effect of this scatter on the POD/CL estimates, one set of experiments was simulated in which it was assumed that there was no scatter in individual crack detection probabilities about the POD(a) function (that is, $S(\epsilon) = 0$).

Comparison of Binomial and Regression Analysis Methods

In the initial attempt to compare the binomial and regression analysis methods, the crack size distribution reported for the Have Cracks data were used in the simulation. For this crack length distribution, the binomial methods always failed to yield estimates of the 90/95 and 95/90 crack length values. For this range of crack lengths in the experiment and for the NDE capability implicitly assumed by the POD(a) function, it was extremely unlikely to obtain any crack length interval that would yield sufficient detections to have 95 percent confidence that 90 percent of all cracks of that length would be found. The regression method, however, always produces a POD/CL value, albeit extremely large in some cases.

When the larger cracks were used in the NDE simulation experiments, the OPM (the recommended binomial method [6]) yielded 90/95 limits in 87 percent of the trials and 95/90 limits in 17 percent of the trials. A more detailed method for comparing the POD/CL estimates from the two analysis methods is to compare the observed distributions of the estimates that were obtained during the 100 simulated experiments. Figure 4 shows the observed distributions of the 90/95 estimates for the regression and OPM analysis methods. In the figure, the vertical scale gives the percentage of the 90/95 estimates that were less than the indicated crack length. For example, using the regression analysis, 50 percent of the 90/95 estimates were less than 40 mm, while using the OPM analysis, 50 percent of the 90/95 estimates were less than 65 mm.

The distributions of Fig. 4 indicate that the regression estimates are more precise in that they have far less variability and that they are generally closer to the "true" 90th percentile crack length of 20.1 mm. Both analysis methods produced estimates that were always above the "true" value but the lack of reproducibility in either of the estimates casts doubts on their usefulness. For

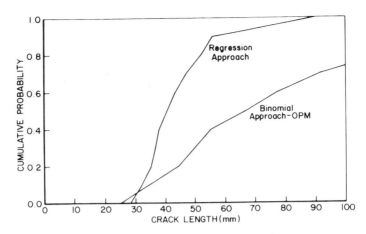

FIG. 4—*Observed distributions of 90/95 estimates for regression and OPM analysis methods.*

example, assume the distributions are representative of the scatter in the estimates from a fixed but unknown set of conditions. If a single NDE experiment of 400 inspections is to be performed and the data analyzed using the regression approach, there would be a 20 percent chance that the 90/95 estimate would be greater than 50 mm. This much potential scatter could greatly influence inspection schedules or risk analysis if the 90/95 value were to be used as a_{NDE}.

Comparison of POD/CL Limits

The choice of the POD/CL combination to be used in defining the capability of an NDE system has been rather arbitrarily defined as 90/95. To evaluate various choices from the viewpoint of their estimates in an NDE experiment, the crack lengths corresponding to several combinations of POD and confidence level were calculated for each simulated experiment. The statistical properties of these POD/CL limits under fixed conditions provided considerable insight into the practical usefulness of various combinations.

Table 1 presents the mean ($\bar{X}$), standard deviation (S), and coefficient of variation ($CV = 100 \, S/\bar{X}$) of the estimates of POD/CL limits obtained from the regression analysis. The statistics are based on one sample of 100 simulated "experiments." The observed percentage of POD/CL values greater than the true POD crack length, a_p, was always greater than or equal to the theoretical CL value. Thus, the calculated POD/CL values were conservative.

The coefficient of variation column of the table displays the estimation precision decreases rapidly with increasing POD and with increasing level of confidence. Further, the average of the calculated POD/CL limits increases as the degree of confidence increases as would be expected. The combination of these

TABLE 1—*Means, standard deviations, and coefficients of variation for combinations of POD/CL.*

POD	CL	a_p, mm[a]	Log Odds-Regression		
			$\bar{X}$, mm	S, mm	CV, %
0.5	0.5	5.5	5.6	0.4	7
	0.9		6.1	0.5	8
	0.95		6.4	0.5	8
	0.99		6.9	0.7	10
0.75	0.5	10.5	11.4	1.6	14
	0.9		14.2	3.0	21
	0.95		15.5	4.1	26
	0.99		21.1	12.4	59
0.9	0.5	20.1	24.1	6.5	27
	0.9		36.6	17.8	49
	0.95		45.0	31.8	71
	0.99		122.4	433.8	354
0.95	0.5	31.1	40.4	15.5	38
	0.9		72.6	58.2	80
	0.95		99.8	131.3	132
	0.99		747.3	**[b]	691
0.99	0.5	82.2	130.8	90.9	69
	0.9		383.3	838.8	216
	0.95		802.9	**	409
	0.99		**	**	969

[a] a_p indicates crack length for which 100 POD percent of cracks would be detected.
[b] ** indicates value was larger than 1000.

facts indicates that considerable real scatter is present in the estimate of NDE capability at high values of POD and confidence.

As another example of the effect of scatter in the estimates, assume an a_{NDE} value is to be determined and the data will be analyzed using the regression approach. Figure 5 displays the distribution of potential estimates of a_{NDE} if a_{NDE} is defined as either the 90/95 or 95/90 limit. The result of the future experiment is equivalent to drawing a number at random from either of these cumulative distribution functions. For this NDE capability, the "true" 90th percentile of POD is 21.1 mm, but there is a 50 percent chance that the 90/95 estimate will exceed 40 mm, a 25 percent chance that the estimate will exceed 48 mm, and 10 percent chance that the estimate will exceed 55 mm. These values can be read from the 90/95 curve of Fig. 5. Similarly, the "true" 95th percentile is 31.1 mm while there is a 50 percent chance that the 95/90 estimate will exceed 75 mm and a 22 percent chance the estimate will exceed 100 mm.

In general, the scatter in the estimates is sufficiently large as to cast considerable doubt on the usefulness of any single POD/CL limit if the POD is 0.9 or

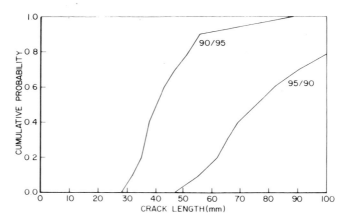

FIG. 5—*Observed distributions of 90/95 and 95/90 estimates from regression analysis method.*

greater and the level of confidence is 0.9 or greater. It should be noted that the scatter in the limits gives rise to excessively large estimates of a_{NDE} and the estimates are conservative. However, the degree of conservativeness would be unknown in a particular application.

Influence of Crack Sizes in NDE Experiments

To compare the binomial and regression model methods of analysis, it was necessary to introduce a distribution of long cracks for the representative "specimens" of the simulation. When the resulting POD/CL estimates from the long cracks simulation were compared to those of the short cracks (that is, the crack sizes of the Have Cracks data) simulations, it was observed that significantly different distributions resulted, Fig. 6. In all cases considered, the long crack experiments had less scatter in the POD/CL estimates than did the experiments with short cracks.

While the effect of specimen crack size distribution has not been sufficiently determined, these comparisons definitely indicate that the sizes of the cracks in a NDE capability demonstration program should be considered as an experimental control to the extent possible.

Influence of Scatter in Detection Probabilities at a Crack Length

In an effort to isolate the causes of the large degree of scatter in the estimates of the POD/CL limits, it was postulated that this scatter could be caused by the relatively poor correlation of crack detection with crack length that was present in the Have Cracks data. To test this hypothesis, NDE simulations were performed with standard error of deviations about the POD function

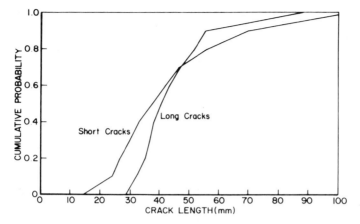

FIG. 6—*Observed distributions of 90/95 estimates from regression analysis of long and short crack experiments.*

reduced to zero. That is, it was assumed that all cracks of a given length had exactly the crack detection probability as given by the POD(a) function. The relative degrees of scatter and the resulting distributions of 90/95 limits are shown in Fig. 7. In Fig. 7 (*left*), the outside bands are 95 percent confidence limits on the detection probabilities as derived from the Have Cracks data; that is, these are 95 percent confidence bounds for the individual data points of Fig. 2. The center curve is the assumed POD(a) function. Figure 7 (*right*) presents the cumulative distributions of 90/95 estimates from the normal scatter and no scatter simulated experiments. Reducing the scatter about the

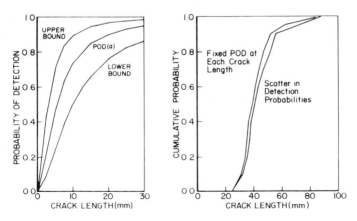

FIG. 7—*Evaluation of scatter in detection probability;* (left) *correlation of detection probability with crack length, and* (right) *distributions of 90/95 estimates.*

POD(a) function to zero, reduced the variability of the POD/CL estimates but not by a practically significant amount.

Discussion

Analysis of the results of the simulated NDE experiments lead to three major conclusions:

1. The large degree of variability in the POD/CL crack length estimates for POD values of 0.9 and greater indicates that such estimates are not reproducible if an NDE capability experiment of the sample size simulated herein would be repeated.
2. Both the magnitude and scatter in the POD/CL estimates are significantly influenced by the crack sizes in the experiment.
3. The variability of the POD/CL estimates is not primarily due to the lack of a strong correlation of detection probabilities of individual cracks with crack length.

These conclusions imply that the scatter in the POD/CL estimates is inherent to the analysis procedure and the sample size but only indirectly to the NDE capability. The simulated experiments of this study were based on inspections of 400 cracks. These would be considered as large experiments due to the difficulty of obtaining representative, cracked structural specimens. Thus, while increasing the sample size would increase the precision of the POD/CL estimates, the very large sample sizes required for a significant decrease in scatter would not be practical.

To further explore the instability of crack length estimates corresponding to high POD values, consider the shape of the model for POD as a function of crack length. Available data from NDE reliability experiments indicate that at least some of the longer length cracks fail to be detected on occasion. Realistic POD models will account for these misses by asymptotically approaching one. Simple geometric considerations lead to the conclusion that estimates of crack lengths corresponding to POD values in the flat portion of the curve are very sensitive to "errors" in the POD value (see Fig. 2). Since the POD value is being estimated statistically, very large sample sizes would be required to reduce the "error" in the POD estimate to yield a precise corresponding crack length.

It is theoretically possible to have an NDE system for which the slope of the POD curve is sufficiently steep, that reasonably precise estimates of the crack length corresponding to a POD of 0.90 or 0.95 can be obtained. Such POD curves have not yet been shown to occur in field applications since human factors as well as inspection hardware influence the capability of the system. Even if such a system were available, however, attempts to characterize it in terms of higher POD levels (say 0.99 or 0.999) would lead to the same lack of precision in the POD/CL estimates.

Conclusions

The following conclusions are based on analysis and the results of the computer simulations of NDE capability experiments that were performed during this study.

1. All cracks of the same length do not have the same detection probability. The strength of the correlation between detection probability and crack length depends on the NDE system (including inspection environment) and the structural elements being inspected.

2. The probability of detection (POD) as a function of crack length for the total population of structural details is a curve through the averages of the detection probabilities for all cracks of the same length. Such a curve is known as a regression function.

3. The log odds model was acceptable as a regression function of the crack detection probabilities that were observed during the Have Cracks program.

4. Confidence limits can be placed on the POD function using regression analysis methods.

5. Given an acceptable model for the regression function, the regression estimates of NDE capability expressed in terms of a confidence limit on a high probability of detection value (that is, a POD/CL value) are superior to those derived using binomial distribution theory. The regression estimates are closer to the true POD, exhibit less scatter in the distribution of the estimates, and, contrary to binomial methods, always provide an estimate of the desired limit.

6. For the NDE experiments simulated, the magnitude and scatter of the POD/CL values are significantly influenced by the crack sizes employed in the NDE capability experiment.

7. For the NDE experiments simulated, the degree of scatter of the detection probabilities of individual cracks about the POD function has only a secondary effect on the scatter in the POD/CL estimates.

8. For the NDE experiments simulated, single number characterizations of NDE capability expressed in terms of a probability of detection and a confidence level (POD/CL) display a degree of scatter (that is, nonreproducibility) that make these characterizations of limited practical use in the evaluation of NDE systems.

Acknowledgment

This work was supported by the Air Force Materials Laboratory (AFWAL/ML) under Contract F33615-80-C-5140.

References

[1] "Aircraft Structural Integrity Program, Airplane Requirements," Military Standard MIL-STD-1530A, 1975.
[2] "Airplane Damage Tolerance Requirements," Military Specifications MIL-A-83444, 1974.

[3] Gallagher, J. P., Grandt, A. F., Jr., and Crane, R. L., *Journal of Aircraft,* Vol. 15, July 1978, pp. 435-442.
[4] Coffin, M. D. and Tiffany, C. F., *Journal of Aircraft,* Vol. 13, Feb. 1976, pp. 93-98.
[5] Packman, P. G., Klima, S. J., Davies, R. L., Malpani, J., Moyzis, J., Walker, W., Yee, B. G. W., and Johnson, D. P., "Reliability of Flaw Detection by Nondestructive Inspection," *ASM Metal Handbook,* Vol. 11, 8th ed., Metals Park, Ohio, 1976, pp. 214-224.
[6] Yee, B. G. W., Chang, F. H., Couchman, J. C., Lemon, G. H., and Packman, P. G., "Assessment of NDE Reliability Data," NASA CR-134991, National Aeronautics and Space Administration, Lewis Research Center, Cleveland, Ohio, 1976.
[7] Lewis, W. H., Dodd, B. D., Sproat, W. H., and Hamilton, J. M., "Reliability of Nondestructive Inspections—Final Report," Report No. SA-ALC/MEE 76-6-38-1, U.S. Air Force, San Antonio Air Logistics Center, Kelly Air Force Base, Tex., 1978.
[8] Natrella, M. G., *Experimental Statistics,* Handbook 91, National Bureau of Standards, 1963.
[9] Cox, D. R., *The Analysis of Binary Data,* Methuen and Co., London, 1970.
[10] Finney, D. J., *Statistical Method in Biological Assay,* Hafner, New York, 1964.
[11] Berens, A. P. and Hovey, P. W., "Evaluation of NDE Reliability Characterization," AFWAL-TR-81-4160, Air Force Wright Aeronautical Laboratories, Wright-Patterson Air Force Base, Ohio, Nov. 1981.
[12] Dixon, W. J. and Massey, F. J., Jr., *Introduction to Statistical Analysis,* McGraw-Hill, New York, 1957.

Statistical Aspects of Fatigue

D. F. Ostergaard[1] *and B. M. Hillberry*[2]

Characterization of the Variability in Fatigue Crack Propagation Data

REFERENCE: Ostergaard, D. F. and Hillberry, B. M., **"Characterization of the Variability in Fatigue Crack Propagation Data,"** *Probabilistic Fracture Mechanics and Fatigue Methods: Applications for Structural Design and Maintenance, ASTM STP 798*, J. M. Bloom and J. C. Ekvall, Eds., American Society for Testing and Materials, 1983, pp. 97-115.

ABSTRACT: The variability in fatigue crack propagation is studied through the analysis of 68 sets of replica constant amplitude fatigue crack growth data. Each of the a versus N data sets are fit to the integral of a crack growth rate equation through a finite integral optimization routine. The routine determines the best fit crack growth parameters of the crack growth rate equation. Through characterization of the variability in the crack growth parameters and utilizing statistical theory, the variability in the a versus N data is predicted. The finite integral method of obtaining the crack growth parameters was shown to very accurately reproduce the a versus N data. The variability in the crack growth parameters showed a strong linear trend, and a regression analysis on the parameters provided information to accurately predict the variability in the fatigue crack propagation data.

KEY WORDS: fatigue (materials), fatigue crack growth, statistical analysis, data analysis, variability, probabilistic fracture mechanics, fracture mechanics

The current approach in characterizing subcritical fatigue crack propagation data is to process the a versus N data into a differential form, da/dN, and present it as a function of the applied stress intensity, K (or ΔK). Presently, there are many differentiation schemes used to obtain the required da/dN data. The variety of methods arises from the fact that the derivative of the function $a = f(N)$ is not completely defined but instead is approximated at discrete points from the a versus N data. Some of these methods analyze the data on a point by point basis while others analyze the data over a series of data points [1].[3]

---

[1]Engineer, Alcoa Technical Center, Alcoa Center, Pa. 15069.
[2]Professor, School of Mechanical Engineering, Purdue University, West Lafayette, Ind. 47907.
[3]The italic numbers in brackets refer to the list of references appended to this paper.

An accurate crack growth rate equation describing the fatigue process of a test specimen will, upon integration, reproduce the original a versus N data. The current method of differentiating the a versus N data to obtain the crack growth rate has been shown to introduce considerable error depending on the particular differentiation scheme utilized [2] and the Δa measurement increment [3]. It has been shown that regression analysis of the differentiated data followed by subsequent integration of the resulting crack growth rate equations leads to inconsistent reproduction of the actual data [1,4].

The discrepancies arising from this approach appear to result from using derived da/dN data to reproduce a versus N data. The extensive regression and curve fitting of the da/dN data has been shown to mathematically characterize the log da/dN versus log ΔK data, but subsequent integration of the resulting equation has shown that the models do not necessarily represent the fatigue crack growth (a versus N). Instead of using derived da/dN data to obtain the parameters of the crack growth rate equation, use of the a versus N data to obtain the parameters would provide a direct correlation between the growth equation and the raw data. Hence, integration of the crack growth rate equation using this approach would reproduce the a versus N data more closely.

In this investigation, a versus N data sets collected by Virkler et al [2] from identical specimens under identical load conditions are fit to the integral of a crack growth rate equation. Each data set is individually fit to the crack growth rate equation by determining crack growth parameters in the equation through a finite integral optimization routine. The routine uses solely the a versus N data to obtain the crack growth parameters.

If a single crack growth rate equation accurately represents individual test specimen fatigue characteristics under identical loading conditions, then the variability in the fatigue process can be represented by characterizing the variability in the parameters of the growth rate equation. This is one of the probabilistic approaches to modeling the variability in the fatigue process considered by Kozin and Bogdanoff [5].

In the study presented here, the preceding method for determining the parameters of the growth rate equation was applied to each of Virkler's 68 tests to determine the variability in the crack growth rate parameters. Using these results, a Monte-Carlo type prediction scheme was used to predict the 68 sets of a versus N that was then compared with the original a versus N data.

Experimental Data

The experimental data utilized in this study were obtained by Virkler et al [2]. Virkler investigated both the variability in the number of cycles, N, to grow a crack a specified distance and the variability in the crack growth rate, da/dN. He conducted 68 replicate tests and examined the distribution of N and also da/dN using several differentiation methods. In addition, he did a comparative analysis to determine the differentiation method and the statistical distributions that best represented the data.

The test specimens used by Virkler in his investigation were 2.54-mm (0.10-in.) thick, center crack panels of 2024-T3 aluminum alloy cut from a single lot of material. The specimens were 152 mm (6 in.) wide, and 55.8 mm (22 in.) long. The specimens were numbered as received and tested in a random order. To ensure a dry atmosphere at the crack tip, the entire expected crack path on both sides of the specimen was sealed with clear polyethylene enclosing a silica gel dessicant.

The test machine utilized by Virkler was a 20 Kip electro-hydraulic, closed-loop system operated in load control. He used a digital cycle counter to count the number of applied load cycles. Virkler measured the crack length with a ×150 zoom stereo microscope mounted on a horizontal and vertical digital traversing system. A digital resolver on the traverse provided a direct digital output of the crack length. The output from the digital resolver and cycle counter were connected to a printer that permitted direct, simultaneous recording of the crack length and number of cycles. The crack tip was illuminated by a strobe light synchronized to trigger when the crack was most fully open in each cycle. The data were recorded at constant Δa increments by advancing the microscope traverse 0.2 mm and pressing the printer push button when the tip of the crack reached the cross-hair in the microscope.

The crack was initiated from an electro-discharge-machined (EDM) slot and grown to a one-half crack length of 9.0 mm. The crack initiation followed a prescribed load reduction schedule so as to ensure that no load effects were present at the start of data acquisition. Only one side of the crack was measured.

Virkler ran the tests under constant load amplitude condition with the load controlled to within 0.2 percent. The test conditions are listed here.

$$P_{min} = 476 \text{ Kg } (1050 \text{ lb})$$
$$P_{max} = 238 \text{ Kg } (5250 \text{ lb})$$
$$R = 0.20$$

The crack was grown from 9.0 mm to 49.8 mm continuously. Three different Δa increments were used to record the data due to the faster growth rates in the latter region. The a versus N data collected from Virkler's 68 tests are shown in Fig. 1.

Crack Growth Rate Equation

An analysis of log-log plots of da/dN versus ΔK data from Virkler's study indicated that the growth rate curve could be divided into two regions. A linear region on the log-log plots (Region II) was noted above a crack length of approximately 28 mm. Below this the log-log data followed an asymptotic decay (Region I). Region III depicted by nonlinear behavior [6] commonly noticed in 2024-T3 aluminum alloy data in the high stress intensity range was not ob-

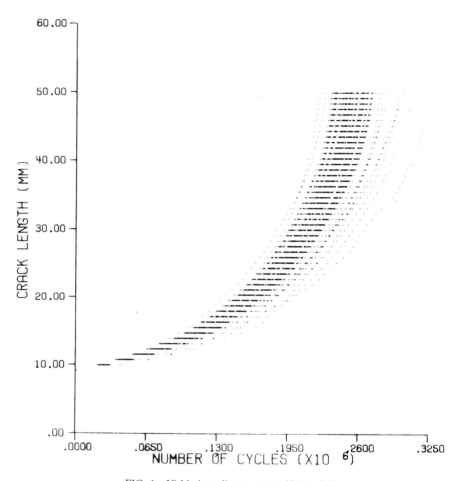

FIG. 1—*Virkler's replicate* a *versus* N *data [1].*

served in Virkler's [2] data. This was due to the low stress intensity range in which the data were collected and therefore this third region behavior was neglected in the formulation of the equation.

The growth rate in Region II can be described by the Paris law

$$\frac{da}{dN} = C(\Delta K)^m \qquad (1)$$

where C and m are empirically derived constants. The asymptotic decay in Region I can be attributed to a greater decrease in growth rate in the low ΔK region converging to a threshold value, ΔK_{th}, below which the crack will not grow [7]. For an analytical relationship, the limitation implies

$$\frac{da}{dN} \to 0 \quad \text{if } \Delta K - \Delta K_{th} \to 0$$

This relationship is easily satisfied by modifying Eq. 1 to the form

$$\frac{da}{dN} = C(\Delta K - \Delta K_{th})^m \tag{2}$$

that is the form of the crack growth rate equation used in this investigation for the analysis of Virkler's data. The value of the threshold stress intensity parameter, ΔK_{th}, utilized in the crack growth rate equation was 3.0 MNm$^{-3/2}$ [8].

Finite Integral Optimization Routine

The values for the crack growth parameters, C and m, in Eq 2 are obtained for a given test specimen by using the raw a versus N data directly. To obtain the parameters directly from the raw data, Eq 2 is rearranged to the form

$$\int_{a_0}^{a} \frac{da}{C(\Delta K - \Delta K_{th})^m} = \int_0^N dn = N \tag{3}$$

where a_0 is the initial crack length, and a and N are a corresponding set of crack length and cycle count data. A set of crack growth rate parameters, C and m, are assumed, and Eq 3 is numerically integrated through finite crack length levels. A new set of parameters, C and m, are selected, and Eq 3 is integrated again. This is repeated until an optimum set of C and m is obtained that results in the closest fit to the raw data.

In this study, Eq 3 was integrated at a constant Δa increment of 0.8 mm. An error term, E_k, was recorded for each chosen set of crack growth parameters and was defined as

$$E_k = \sum_{i=1}^{A_t} (N_{ai} - N_{ri})^2 \tag{4}$$

where N_{ai} is the cumulative cycle count from experimentally recorded data, N_{ri} is the cumulative cycle count from the numerical integration, and A_t is the total number of crack length levels. The optimum value of the crack growth parameters was determined as the set that produced the minimum error, E_k. The optimization was performed using a nongradient based minimization algorithm developed by Hooke and Jeeves [9].

Each of Virkler's [2] a versus N data sets were individually fit to Eq 3 resulting in 68 sets of reproduced a versus N data. Figure 2 shows a typical fit of a reproduced data set for Test 2.

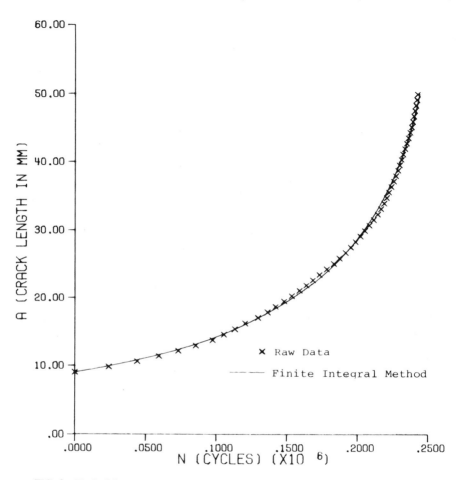

FIG. 2—*Typical fit of a versus N data from integration of crack growth equation (Eq 2).*

An error analysis was performed that compared the raw data with the reproduced data from the crack growth rate model. At each crack length level the percent difference between the reproduced N and the actual N was determined from the equation

$$E_r = \frac{N_{ri} - N_{ai}}{N_{ai}} \times 100 \qquad (5)$$

The error analysis was performed on all 68 raw data sets and their corresponding reproduced data sets. Figure 3 is a plot of the percent error versus crack length for all 68 tests at each crack length level. The error analysis shows that

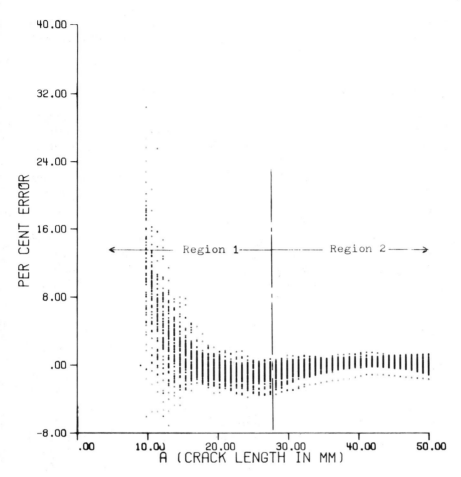

FIG. 3—*Percent error in reproducing Virkler's [1] original a versus N data using the crack growth equation (Eq 2) fit to each replicate test.*

the reproduced data from the crack growth rate model provided an excellent fit to the original raw data in all cases with only minor discrepancies at the beginning of the data collection region. On the average, the error in reproducing N was less than one percent from 14.6 to 49.8 mm.

A sensitivity study was done to investigate the effect of the ΔK_{th} parameter on crack growth Parameters C and m and on the resulting error in predicting the crack growth data as determined in Eq 5. A single set of Virkler's a versus N data was fit to Eq 2 through the finite integral routine and parameters C and m determined for ΔK_{th} values varying ±10 percent of the chosen value of 3.0 MNm$^{-3/2}$. Results showed that Parameter m remained unchanged and that variation in the ΔK_{th} parameter was entirely absorbed in Parameter C. A

10 percent variation in ΔK_{th} resulted in about a 2 percent variation in Parameter C. The error analysis showed identical values of E_r for all three cases ($\Delta K_{th} = 3.0$, $\Delta K_{th} = 3.0 \pm 10\%$).

Finite Integral Method versus Differentiation Methods

A comparative analysis was performed to investigate the accuracy between the integral method of obtaining crack growth parameters for a growth rate equation and the current method of regression analysis on differentiated a versus N data in obtaining crack growth parameters.

The data obtained for this analysis were from a round-robin testing program that investigated the variability and bias associated with fatigue crack growth rate testing [10]. The test specimens were 6.35 mm (0.25 in.) thick compact tension specimens (WOL type) of 10Ni-8CO-1Mo steel. The data were obtained from several different laboratories. Each laboratory supplied for analysis the a versus N data, the fatigue crack growth rate da/dN as a function of the stress intensity parameter ΔK, and Parameters C and m of a Paris law fit to a log-log plot of da/dN versus ΔK data (Eq 1). Each laboratory was free to select the method of determining the crack growth rates for their data.

Eleven sets of data from the round-robin study were selected for investigation. The data sets were selected randomly from each group that had been analyzed with a different differentiation scheme. A summary of the test identification numbers, crack growth rate determination methods, and the C and m parameters as reported by the round-robin study [10] are shown in Table 1. The C and m parameters in the table were obtained from a least squares regression analysis of the log da/dN versus ΔK data.

The finite integral routine was applied to each data set using the same crack growth rate equation as was used in the round-robin study (Eq 1). The finite integral routine determined the C and m parameters for each test. These results are included in Table 1.

To investigate the accuracy in reproducing the a versus N data for the two methods, the crack growth rate equation was integrated using the derived C and m parameters, and the resulting reproduced a versus N data sets were compared with the original data. At each crack length level for each test, the percent error in reproducing the original N data was calculated using Eq 5. An average percent error from all cycle count reproduction errors at each crack length level was determined for each test and is shown in Table 1 for the two methods.

The error analysis clearly indicates that in each case the finite integral optimization routine was superior in providing the best representative crack growth parameters for reproducing the a versus N data.

Figures 4 and 5 compare the regression fit and the finite integral fit to one of the test data sets. Figure 4 shows the da/dN data as obtained using a secant method of differentiation with a least squares regression fit and the finite inte-

TABLE 1—*Summary of the crack growth rate equation parameters and error terms for the WOL type specimen data analyzed.*

Test Identification	Differentiation Method	Regression Analysis			Finite Integral Analysis		
		$C (\times 10^{-9})$	m	Average Reproduction Error, %	$C (\times 10^{-9})$	m	Average Reproduction Error, %
1-91A	Secant	12.16	1.89	11.56	4.00	2.21	1.85
1-45A	method	15.77	1.84	12.68	4.87	2.17	3.70
2-36A		18.01	1.71	10.36	4.84	2.10	1.94
17-68A	Modified	12.60	1.98	1.27	17.14	1.88	0.52
17-68B	difference	8.97	2.06	2.86	5.36	2.22	2.20
17-86A	method	4.83	2.23	5.58	2.37	2.45	1.46
14-100A	Graphical	22.41	1.78	2.26	22.11	1.78	1.29
14-100B	method	23.89	1.75	1.35	16.07	1.88	0.82
14-106A		23.32	1.77	3.53	10.78	2.01	1.66
7-70A	Total	3.62	2.31	3.85	6.21	2.15	1.03
7-70B	polynomial method	5.25	2.21	3.63	6.73	2.12	0.94

gral fit. Note that the finite integral fit has the appearance of being more sensitive to the lower ΔK data. This is explained by the fact that Parameters C and m chosen by the integral method were fit from the a versus N data, and it turns out that about 75 percent of the life was spent below a ΔK of 36 ksi $\sqrt{\text{in.}}$ Figure 5 shows a comparison of the raw a versus N data and the a versus N data obtained by integrating the growth rate curves in Fig. 4. Note the superiority of the finite integral fit over the regression method in reproducing the a versus N data. (Note: The finite integral fit to this set of data can be substantially improved by using a more appropriate crack growth rate equation, such as Eq 3.)

Predicting the Variability of Crack Growth Data from Crack Growth Parameters

In investigating the relationship between the 68 sets of crack growth parameters, it was found that a log-log plot of C and m could be well represented by a straight line as shown in Fig. 6. From the regression analysis with log m the independent parameter

$$\widehat{\log C} = b_0 + b_1 \log m \tag{6}$$

where

$\widehat{\log C}$ = fitted value of log C, and
b_0 and b_1 = least squares estimated regression parameters.

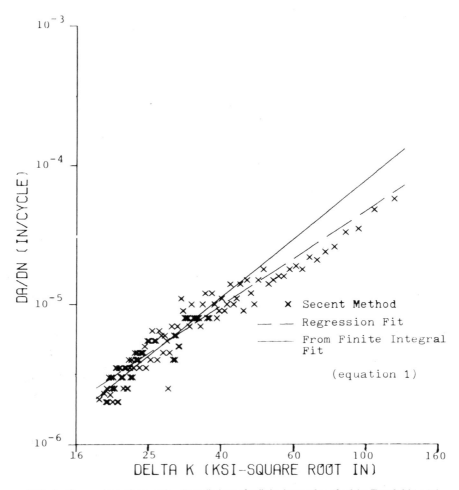

FIG. 4—*Comparison of regression fit to fit from the finite integral method for Test 2-36A of the round-robin study [10].*

The values determined for b_0 and b_1 were

$$b_0 = -5.7792$$

$$b_1 = -4.6150$$

The significance of this rather strong relationship between Parameters C and m was further investigated through the regression line parameter combinations. The mean a versus N curve obtained from Virkler's [2] raw data was plotted against four a versus N curves obtained by integrating Eq 2 with crack growth

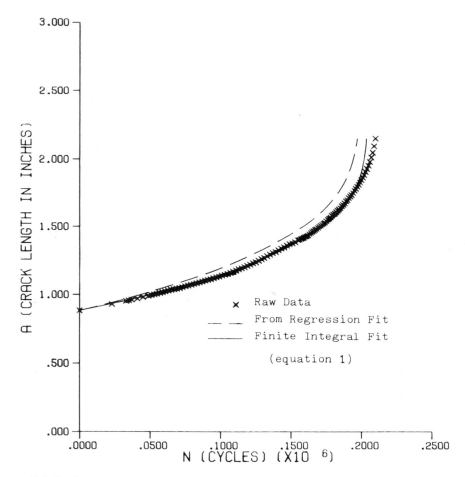

FIG. 5—*Comparison of integrated regression data with finite integral fit for Test 2-36A of the round-robin study [10].*

parameter combinations taken from the regression line in Fig. 6. The resulting plots are shown in Fig. 7. The crack growth parameter combinations taken from the regression line are representative of the range of scatter in the parameter combinations. The resulting a versus N curves in Fig. 7 suggest that the regression line provides a very accurate representation of the combination of crack growth parameters that will reproduce the mean a versus N curve, upon integration in the crack growth rate equation.

Since the regression line in Fig. 6 is representative of the crack growth parameter combinations that reproduce the mean a versus N curve, it was hypothesized that a measure of the deviation of the parameter combinations from the regression line could be used to predict the variability in the raw a versus N data.

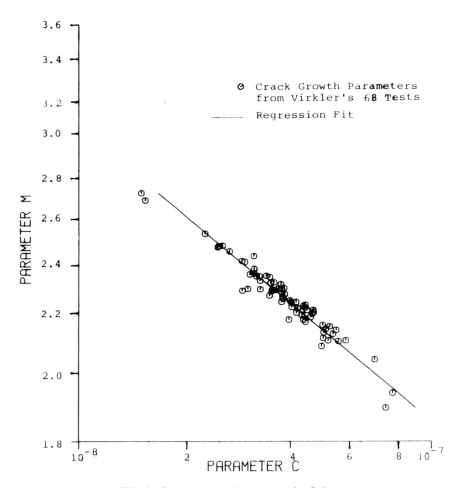

FIG. 6—*Regression fit to log m versus log C data.*

To characterize the deviation a Parameter F was defined as

$$F = \frac{\hat{C}}{C} \quad (7)$$

where $\widehat{\log C}$ is the fitted value of log C from the regression analysis. Taking the logarithm of Eq 7 and rearranging terms gives

$$\log F = -[\log C - \widehat{\log C}] \quad (8)$$

where log F represents the least squares residuals from the regression analysis.

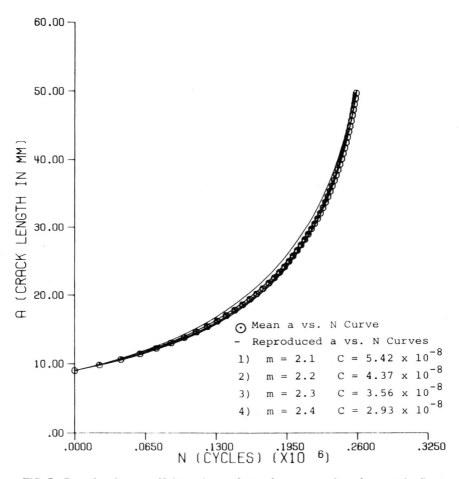

FIG. 7—*Reproduced a versus N data using crack growth parameters from the regression line.*

Since $\Sigma (\log F_i)$ are the residuals, a plot of $\log F$ versus $\log m$ will show no linear trend. The data can be standardized by defining functions A_f and A_m as

$$A_F = \log F/S_F \qquad (9)$$

$$A_m = (\log m - \bar{m})/S_m \qquad (10)$$

where S_f and S_m are the standard deviations of Parameter F and Parameter m data, respectively. A plot of these values shown in Fig. 8 reveals no other noticeable trend between $\log F$ and $\log m$. Thus, there is not enough evidence to suggest that F and m are not independent. Based on this, independency between F and m is assumed in the prediction of the a versus N data.

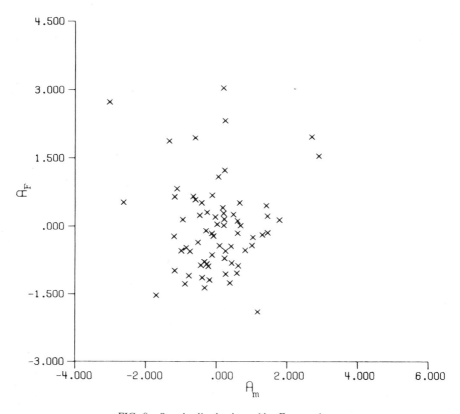

FIG. 8—*Standardized values of log F versus log m.*

Having obtained values for Parameter m from the finite integral optimization routine and values of F from the regression analysis on the log m versus log C data, the variability in these parameters were characterized by fitting a statistical distribution to the parameter data. This was done utilizing program CGRDDP developed by Virkler [2] that fits six statistical distribution functions to a set of data. By using goodness of fit criteria, the program determines the best-fit statistical distribution. The analysis showed that the two-parameter log normal distribution best characterized the variability in Parameter m while the three-parameter log normal distribution best characterized the variability in Parameter F. The density function for the three-parameter log normal distribution for characterizing the variability in Parameter F is given by [11]

$$F(\chi) = \frac{1}{(\chi - \tau)\sqrt{2\pi\beta}} \exp\left(-\frac{1}{2\beta}\{\log_{10}(\chi - \tau) - \mu\}^2\right) \quad (11)$$

where the three parameters of the distribution were estimated as

$\tau = 0.807$,
$\mu = -0.737$, and
$\beta = 0.025$.

Having characterized the variability of Parameters F and m and assuming these two parameters were independent, a Monte-Carlo type prediction scheme was utilized to predict the raw a versus N data. The two-variable prediction scheme to obtain a predicted set of a versus N data was as follows.

1. Randomly choose a value of m from its statistical distribution (m_i).
2. Randomly choose a value of F from its statistical distribution (F_i).
3. Solve for Parameter C_i).
4. Integrate Eq 2 using Parameters C_i and m_i to obtain a predicted set of a versus N data.

The procedure was repeated 68 times, thus predicting 68 sets of a versus N data. Figure 9 is a graphical plot of the predicted a versus N data. Figure 9 may be compared with the original raw data shown in Fig. 1.

It was anticipated that the variability in Parameter F played a more significant role with respect to the variability in the a versus N data than did the variability in Parameter m. To check this, the prediction scheme was repeated, however, Parameter m was held constant at its mean value of 2.27 (single-variable prediction scheme). This effectively reduced the unknowns to only Parameter F. The resulting 68 predicted a versus N data sets are shown in Fig. 10.

To determine if the variability in the fatigue data is described by the prediction method, the Kolmogorov-Smirnov test was used. This test determines quantitatively if the predicted N and the experimentally determined N have the same distribution. The Kolmogorov-Smirnov test was applied at five different crack lengths of approximately equal increments. At each crack length, the cumulative distribution of the 68 predicted values of N were compared with the cumulative distribution of the actual values of N by calculating the maximum positive and negative differences (based on a scale of 0.0 to 1.0).

The test statistic, D, then is the absolute maximum difference chosen from these two values. Tables 2 and 3 list the results of the Kolmogorov-Smirnov test for both prediction schemes.

As a test for goodness of fit between the predicted and actual sets of N data at the five crack lengths, the D statistic values can be compared with critical values corresponding to various significance levels shown in Table 4 [12]. Using a significance level of 0.10, the corresponding critical value of the D statistic is 0.15. The results displayed in Tables 2 and 3 show all values of the D statistic below 0.15, indicating that the predicted cumulative N distribution would fall within the 90 percent confidence bands of the actual cumulative N distribution. Thus, at the 10 percent level, the hypothesis that the predicted and actual N data are drawn from the same population distribution is acceptable.

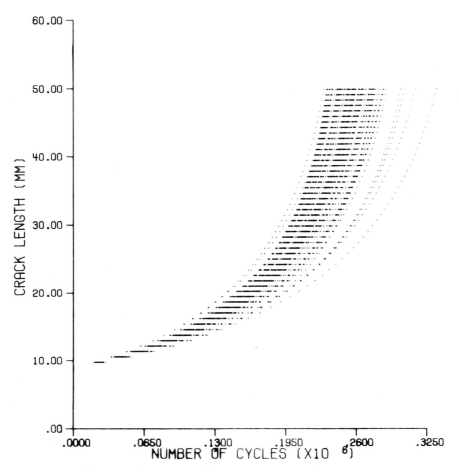

FIG. 9—*Predicted* a *versus* N *data, two-variable prediction scheme.*

The two-variable prediction scheme showed slightly better correlation with the raw data than the single-variable prediction scheme. The two-parameter approach more accurately represents the *a* versus *N* data due to the variability allowed for in both Parameters C and m, however, the single parameter approach, which only allows for variability in Parameter C, does give very good results in predicting the variability in the data and requires only knowledge of Parameter F.

Conclusions

A finite integral optimization routine was developed that fit a set of *a* versus *N* data to the integral of a crack growth rate equation. The routine was ap-

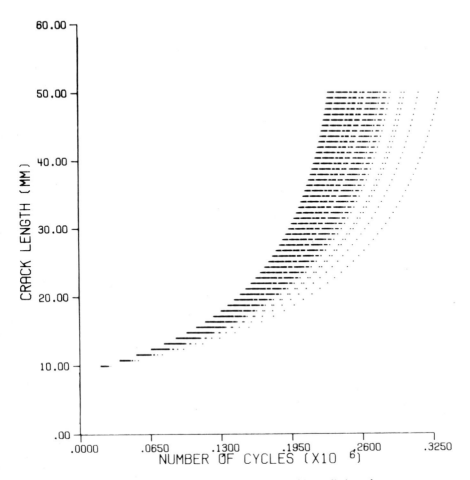

FIG. 10—*Predicted a versus N data, single-variable prediction scheme.*

TABLE 2—*Kolmogorov-Smirnov test results for comparison of predicted N and actual N: two-variable prediction scheme.*

Crack Length, mm	Max (+difference)	Max (−difference)	D Statistic
16.2	0.0588	−0.1471	0.1471
24.2	0.0882	−0.0294	0.0882
32.2	0.1029	−0.1029	0.1029
42.6	0.0735	−0.1029	0.1029
49.8	0.1029	−0.0735	0.1029

TABLE 3—*Kolmogorov-Smirnov test results for comparison of predicted* N *and actual* N: *single-variable prediction scheme.*

Crack Length, mm	Max (+difference)	Max (−difference)	D Statistic
16.2	0.0441	−0.1471	0.1471
24.2	0.1029	−0.0735	0.1029
32.2	0.1029	−0.1324	0.1324
42.6	0.0588	−0.1324	0.1324
49.8	0.0882	−0.1029	0.1029

TABLE 4—*Critical values for the Kolmogorov-Smirnov test for 68 samples.*

Significance level	0.20	0.15	0.10	0.05	0.01
Critical value	0.130	0.138	0.150	0.165	0.198

plied to 68 sets of experimentally obtained a versus N data, and in each case the growth rate equation was shown to reproduce the actual data effectively.

The crack growth Parameters C and m of the growth rate equation obtained for each of the 68 tests were shown to be characteristically related by a least-squares regression line of log m versus log C. Crack growth parameter combinations lying on the regression line were shown to reproduce the mean fatigue curve (a versus N) upon integration in the crack growth rate equation. The least-squares residuals from the regression line of the log m versus log C data were effectively used to characterize the variability in the raw a versus N data through a Monte-Carlo scheme that predicted the raw a versus N data.

References

[1] Hudak, S. J., Jr., Saxena, A., Bucci, R. J., and Malcolm, R. C., "Development of Standard Methods of Testing and Analyzing Fatigue Crack Growth Rate Data," AFML-TR-7840, Air Force Materials Laboratory, May 1978.
[2] Virkler, D. A., Hillberry, B. M., and Goel, P. K., "The Statistical Nature of Fatigue Crack Propagation," AFFDL-TR-78-43, Air Force Flight Dynamics Laboratory, April 1978.
[3] Ostergaard, D. F., Thomas, J. R., and Hillberry, B. M., "Effect of the Δa Increment on Calculating da/dN from a Versus N Data," *Fatigue Crack Growth Measurement and Data Analysis, ASTM STP 738*, American Society for Testing and Materials, 1981.
[4] Pook, L. P., "Basic Statistics of Fatigue Crack Growth," NEL Report No. 595, National Engineering Laboratory, 1975.
[5] Kozin, F. and Bodganoff, J. L., "A Critical Analysis of Some Probabilistic Models of Fatigue Crack Growth," *Engineering Fracture Mechanics*, Vol. 14, 1981.
[6] Forman, R. G., Kearney, V. E., and Engle, R. M., "Numerical Analysis of Crack Propagation in Cyclic-Loaded Structures," *Journal of Basic Engineering*, Transactions of the American Society of Mechanical Engineers, Vol. 89, 1967.
[7] Schivje, J. "Four Lectures on Fatigue Crack Growth," *Engineering Fracture Mechanics*, Vol. 11, 1979.

[8] Crandall, G. M., "Residual Stress Intensity Parameters for Prediction of Delay in Fatigue Crack Propagation," M.S. thesis, Purdue University, West Lafayette, Ind., May 1975.
[9] Hooke, R. and Jeeves, T. A., "Direct Search Solution of Numerical and Statistical Problems," *Journal of the Association of Computing*, Vol. 8, March 1961.
[10] Clark, W. G., Jr., and Hudak, S. J., Jr., "Variability in Fatigue Crack Growth Rate Testing" *Journal of Testing and Evaluation*, Vol. 3, No. 6, 1975.
[11] Atichison, J. and Brown, J. A. C., *The Lognormal Distribution*, Cambridge University Press, Cambridge, U.K., 1957.
[12] Massey, F. J., Jr., *Journal of American Statistical Association*, Vol. 46, 1951, pp. 68–78.
[13] Lindgren, B. W. and McElrath, G. W., *Introduction to Probability and Statistics*, MacMillan Company, New York, 1959, p. 261.

E. K. Walker[1]

Exploratory Study of Crack-Growth-Based Inspection Rationale

REFERENCE: Walker, E. K., "**Exploratory Study of Crack-Growth-Based Inspection Rationale,**" *Probabilistic Fracture Mechanics and Fatigue Methods: Applications for Structural Design and Maintenance, ASTM STP 798*, J. M. Bloom and J. C. Ekvall, Eds., American Society for Testing and Materials, 1983, pp. 116-130.

ABSTRACT: A relatively simple probabilistic model is developed having as its basic elements a log-normal representation of crack initiation and crack growth rates, the probability of crack detection, a crack growth curve, and a safety criterion based on a probability of residual strength equal to or less than limit load capability. Interrelationships between these model elements are explored through selected quantified examples. Some resulting internal relationships and trends are discussed.

KEYWORDS: aircraft, inspections, cracks, probabilities, fatigue (materials), probabilistic fracture mechanics, fatigue crack growth, fracture mechanics

In December 1978, the Federal Aviation Administration (FAA) implemented new Airworthiness Regulations for damage tolerance and durability of commercial transport aircraft primary structure. In May 1981, advisory material was released to assess the conformance of existing commercial transport aircraft with the new regulations and to provide additional inspections to obtain conformance if needed. One of the principal requirements that sets the new regulation apart from its predecessor is the requirement that the aircraft inspection program be structured so that should any crack develop and grow affecting safety of flight structure, it will be detected prior to a reduction in residual strength to limit load capability.

To satisfy this requirement in an economic manner, a rational approach is needed to relate the intervals of safe detectable crack growth to the complexities of inspection programs for modern commercial aircraft. These programs include inspections of the same details in many aircraft of different ages, sampling inspections involving relatively few aircraft, and various mixes of in-

[1]Senior scientist, Research and Development, Lockheed-California Company, Burbank, Calif. 91520.

spection methods each having different repeat intervals and different capabilities for detecting cracks.

This paper develops a relatively simple probibalistic model and explores some of the interrelationships between its principal variables with the objective of gaining understanding of interrelationships between variables affecting crack-growth-based aircraft inspection programs. At present, the problem of how best to quantify a mathematical model for aircraft inspection programs is still unresolved. This is primarily due to a lack of a good data base for judging the merits of one approach as compared to another. With this in mind, the model is developed and quantified with a minimum of discussion and with an eye to simplicity of assumption and ease of visualization.

Model Development

The model consists of four basic elements that will be briefly discussed in the following paragraphs.

1. The distribution in times to reach a given crack size is approximated by a log-normal distribution.
2. All inspection methods are represented by a single curve of probability of detection versus the crack size divided by the size judged to be detectable 50 percent of the time.
3. An estimate of crack growth in a selected structural detail is illustrated by a curve of typical shape without geometric complexities.
4. The safety criterion is an assumed-to-be-acceptable probability of a crack resulting in residual strength equal to or less than limit load capability.

Fleet Cracking Experience

Studies of Horsburg and Senkowski[2] show about a 50 percent confidence that time to cracking in service aircraft can be approximated by a $\sigma_{\log}$ no worse than about 0.12 and about a 90-percent confidence that $\sigma_{\log}$ will be no worse than about 0.2. Dinkeloo and Moran[3] suggest that $\sigma_{\log}$ be between 0.15 and 0.2. To illustrate the model, a $\sigma_{\log}$ of 0.13 has been selected. This $\sigma_{\log}$ will be considered invariant in time. This is equivalent to assuming that all cracking starts from a single initial defect size (or fatigue quality) and that the rate of damage accumulation at any given crack size (damage level) is also a log-normal distribution with $\sigma_{\log} = 0.13$ as shown on Fig. 1. If this fleet experience is

---

[2]Horsburg, H. and Senkowski, P., *Evaluation of $\sigma_{\log}$ Using C-130 Experience*, private communication prepared for C-5A review team, 1972.

[3]Dinkeloo, C. J. and Moran, M. S., *Structural Area Inspection Frequency Evaluation (SAIFE)*, Report No. FAA-RD-78-29, prepared for U.S. Department of Transportation, Federal Aviation Administration, Systems Research and Development Service, Washington, D.C., April 1978.

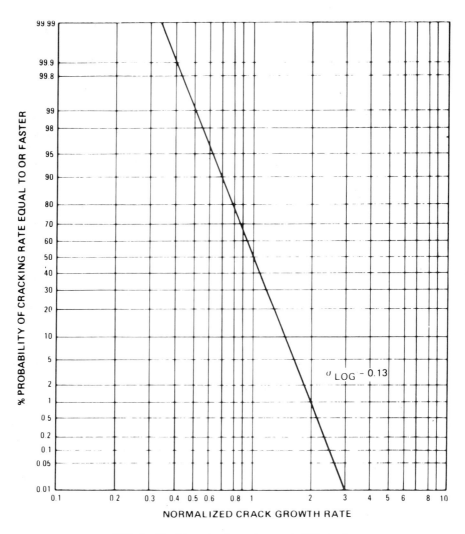

FIG. 1—*Cracking rates for various probability levels.*

estimated using a fracture mechanics approach, it would require selecting a single initial defect and growing a set of cracks from this initial size having at any size, the distribution in cracking rates shown on Fig. 1. The model element is thus the opposite of that being proposed under a current U.S. Air Force program[4]. This latter approach requires a distribution of initial defect

---

[4]*Durability Methods Development*, General Dynamics, Fort Worth Division, Interim Technical Quarterly Report, 15 March 1981 to 15 June 1981, Air Force Flight Dynamics Contract F33615-77-C-3123, 15 June 1981.

sizes from which cracks are grown with all cracks having the same rate at any selected size. The model selected for this study and the model of General Dynamics are obviously both simplified limits of a very complex problem. Both models should lead to approximately the same results when used to interpolate within a body of data. The model selected is the more conservative for extrapolation.

Inspection Effectiveness

Actual inspection effectiveness will vary with structural detail, surface protective system, location, etc. A normalized model for all inspection methods is selected as shown on Fig. 2. The model assumes first that an estimate is made for the size of crack that has a 50 percent chance of being detected by an inspection method considering specifics of detail, location, etc. Once this estimate is made, it is assumed that cracks one half this size will have a 10 percent chance of being detected and two times this size will have a 90 percent chance of being detected. Special considerations, such as a crack growing out from under another member, must be superimposed when necessary. For the illustration, these complexities will be ignored. Figure 2 shows the resulting probabilities versus normalized crack size.

In order to implement the model, three levels of detectability will be examined as they might apply to a specific structural member with crack growth behavior of Fig. 3:

1. general visual; 50 percent chance of detecting a 20-cm crack,
2. directed visual; 50 percent chance of detecting a 5-cm crack, and
3. special directed, nondestructive evaluation (NDE); 50 percent chance of detecting a 1.25-cm crack.

Mean Crack Growth Behavior

The estimation of crack length versus time would normally involve the use of fracture mechanics, supported as available by test data, fractographic evidence, and service experience. Probably the best that can be expected is that the estimation obtained is an average or typical crack growth curve for the case in question. Figure 3 shows an assumed-to-be-typical curve of crack length versus flights for growth from 0.63 to 30 cm. The 0.63 to 30 cm sizes are selected to cover the inspection capabilities assumed in the paragraph entitled Inspection Effectiveness. A 30-cm crack is assumed to give limit load strength. For simplicity of illustration, the curve is shown free of the influences of geometric complexities such as intermediate fastener holes and crack stoppers. It will be assumed that actual crack growth, in the structure being represented, has an equal chance of being faster or slower than shown on Fig. 3 and have a log-normal distribution in rates at any size.

The times to reach practical-to-detect sizes, which will be defined later,

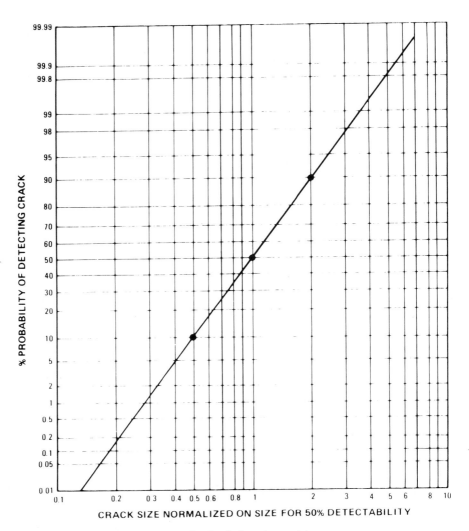

FIG. 2—*Crack detection model.*

will be varied separately. These times could be estimated by either a crack growth analysis from an initial defect or by a fatigue "initiation" analysis or any suitable combination of the two. The separation of estimates of time to reach a practical-to-detect initial size from estimates of time to grow from this initial size to 30 cm is based on several considerations. First, the growth of a crack in the vicinity of fasteners could involve clamp-up, interference, fastener tilt, etc., and thus may not be as reliable to estimate as is the growth through the larger sizes. Second, the division into arbitrary "initiation" and

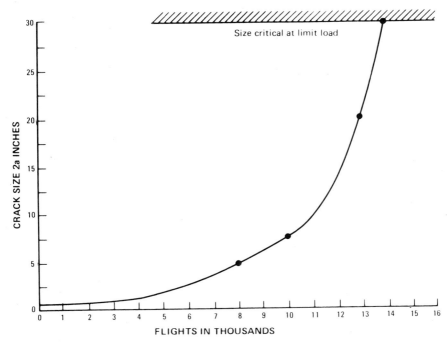

FIG. 3—*Assumed crack growth versus flights.*

propagation segments provides for more flexibility in methodologies. Third, and partially because of the differences in the reliability of estimating the development of small and large cracks, evidence may accumulate during successive inspections to show that initial estimated times to reach detectable crack sizes is in error whereas evidence may not be equally available to substantiate that cracks wil not grow as predicted once they occur.

Safety Criterion

In order to avoid arguments of acceptable probabilities of failure and to remain as close as practical to past deterministic practice, the safety criterion selected will be an assumed-to-be-acceptable probability that a crack resulting in limit load capability or less exists in any aircraft in the fleet as determined by model parameters. As a first guess at what might be rationally acceptable, a probability of 0.01 percent (10^{-4}) is selected.

Model Characteristic

Figure 4 shows a plot on log-normal probability paper combining the fleet cracking experience, crack growth curve, and safety criterion. The typical

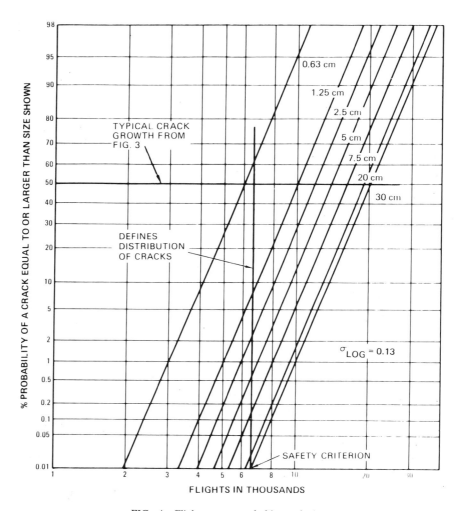

FIG. 4—*Flights versus probable crack sizes.*

crack growth of Fig. 3 is assumed at 50 percent probability. Sloping lines ($\sigma_{\log} = 0.13$) represent the distributions of times required to reach the indicated crack sizes at various probabilities. The vertical line intersecting the size of crack that is critical at limit load (30 cm) at its intersection with a 0.01 percent probability of a crack being equal to or larger than 30 cm, represents the safety criterion. The intersection of this vertical line with various crack sizes describes the probable distribution of crack sizes above 0.63 cm at the time the safety criterion is reached as given by the abscissa (time scale). For any selected crack size from 0.63 to 30 cm, the time to reach or exceed a selected size and probability divided by the time for the same size to be reached or exceeded at the 50 percent probability level provides the ratio of crack growth rates for these two probabilities. For instance, the crack growth rate having 0.01 percent probability of being equaled or exceeded is about 6000 ÷ 2000 or 3 timees that for the 50 percent probability curve. This can be verified by referring to Fig. 1.

To construct Fig. 4, it is also necessary to make a best engineering estimate of the time required to reach a 50 percent probability of having a 0.63-cm crack or larger in the structural detail in question. This is shown as about 6000 flights; an arbitrary selection for illustrative purposes.

It is of interest to examine the relationship in the model between time to first inspection and the average time to reach a practical-to-detect crack size, 1.25 cm in the example. For convenience a practical-to-detect size is defined as the size having a 50 percent chance of being found in a single inspection by a selected inspection method. Crack growth from this size (1.25 cm) to the size for limit load capability is defined as the safe detectable crack growth interval. In the example of Fig. 4, the first inspection, as imposed by the safety criterion, occurs at about 67 percent of the time for the average crack to reach a practical-to-detect size. If the practical-to-detect size had been considered as 20 cm (general visual), the first inspection would be about 35 percent of the time to reach this size. If σ_{log} of 0.2 had been selected, these factors would have been about 33 and 17 percent. Thus, the model provides a rationale for relating the first safety-related inspection relative to the assumed σ_{log}, the average time to reach a practical-to-detect size, and the safe detectable crack growth interval.

This example has not considered the possible use of earlier inspections to detect cracks for economic reasons.

Repeat Inspections

The influence of repeat inspections without finding cracks will now be examined. Table 1 shows a sample set of calculations for a general visual inspection performed at the time the safety criterion is met (inspection threshold) on Fig. 4. The computations were developed as follows. First the probabilities of cracks being equal to or larger than select sizes are taken from Fig. 4 (Table 1, Cols 1 and 2) and approximated by the cumulative probability of cracks between the selected sizes at the mid size for each increment (Cols 3 and 4). Next, the fraction of each size that would typically be missed by the inspection method is determined by entering Fig. 2 at normalized values of crack size obtained by dividing the mid sizes by 20 cm, which is the size judged as having a 50 percent chance of detection with a general visual inspection (Col 5). The cumulative probability of cracks most likely to be remaining of each size range is then calculated by multiplying by the fraction most likely to be missed (Col 6). This is summed to yield the estimated distribution after inspection (Col 7). Next, the growth rate factor (multiple of average rate) is determined from Fig.

TABLE 1—*The effect of a general visual inspection at 6600 flight hours.*

Crack Size, cm	Percent Probability, $\times 10^2$	Percent Probability Increment, $\times 10^2$	Average Size, cm	Fraction Missed	Percent Probability after Inspection, $\times 10^2$	Percent Probability after inspection, $\times 10^2$ (cumulative)	Growth Factor	Equivalent Flights at Mean Rate	Crack Size after 360 Flights, cm
1.25	850	570	1.88	1.00	570	843	1.5	540	1.75
2.5	280	210	3.75	0.998	209	273	1.8	414	3.25
5	70	55	6.25	0.985	54	64	2.1	756	6.50
7.5	15	12	13.75	0.75	9	10	2.4	884	9.25
20	3	2	25	0.32	1	1	2.8	102	30
30	1								

NOTES—

Fraction of cracks detected = $\dfrac{850 - 843}{850} = 0.0082$.

Probability of detecting a crack = $850 \times 0.0082 \times 10^{-2} = 0.07\%$.

1 using the probabilities of these sizes being equaled or exceeded without considering the effect of inspection (Col 8). From Col 7 it can be seen that after inspection, a 20-cm crack has a probability of about 0.01 percent of remaining in the structure. The safety criterion will be met again when a 20-cm crack growing at about 2.8 times the average rate reaches 30 cm. This is the average time for the interval (1000 flights from Fig. 3) divided by 2.8, or 360 flights. For other crack sizes the growth in 360 flights is also obtained. The change in length for a 7.5-cm crack is obtained by advancing from that size on Fig. 3 for a time interval of 2.4 × 360 flights. The growth of other sizes is determined in a similar manner. The sizes after 360 flights are then tabulated (Col 10). The probabilities and sizes after 360 flights (Cols 7 and 10) are plotted on Fig. 5 and should fair smoothly into the estimate of probabilities of small cracks, which are virtually unaffected by the inspection. Columns 7 and 10, plus new cracks becoming of inspectable size determined from Fig. 5, become the basis (Cols 2 and 1) for repeating the procedures of Table 1 to determine the effect of a second inspection. The estimated results of successive general visual inspections are added to Fig. 5 as calculations are made.

The probability of finding a crack of any size during the inspection is estimated and is provided at the bottom of Table 1. A sum of these estimated probabilities for successive inspections on a single aircraft detail is the probability of finding a crack during the series of inspections. The sum may be used as each successive calculation has an appropriately reduced probability of there being a crack as a result of prior inspections.

Tables 2 and 3 show similar calculations for general visual and directed visual inspections. Figure 6 compares the estimated populations at the time a repeat inspection is called for by each method. The populations are approximately the same in the larger sizes most influencing safety.

The preceding inspection rationale takes advantage of both the probability of a crack being in a structure and the probability of detection in satisfying the safety criterion. As such, it is capable of emphasizing a most important point. For inspections of normal effectiveness, structural safety depends to a large extent upon a low probability of cracking.

It should be noted that in each case (Tables 1 through 3), the repeat interval for inspection is about one-third the safe crack growth interval. This is in general conformance to reduction factors commonly used; however, the rationale is different.

Fleet Inspection Programs

Up to this point, the examples and discussions have been concerned with the inspection of an individual structural detail. This same structural detail may exist in similar aircraft of a fleet subject to approximately the same use, or to a variety of uses that can be reduced to equivalent time of a selected baseline use. Where these are the cases, the inspection information gathered dur-

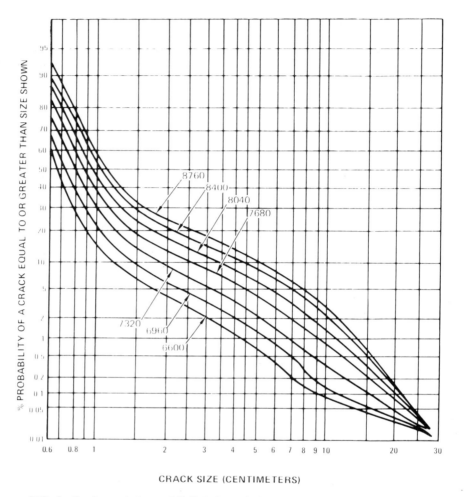

FIG. 5—*Crack populations at 360-flight intervals for repeated general visual inspections.*

ing inspections on aircraft in the fleet can be examined collectively for the purpose of evaluating the estimate of mean time to reach a detectable crack size in a fleet and possibly for evaluating the safe detectable crack growth interval. It is presumed that the safety of individual aircraft with cumulative use beyond that given by the safety criterion is preserved by periodic inspections in the manner previously described. In the model, the purpose of examining fleet inspection data is to assess the reasonableness of the estimated mean crack time history on which the inspections are based. Table 4 shows a hypothetical distribution of aircraft age in flights and the number of directed visual inspections that would have accumulated in accordance with the expected cracking experience of Fig. 4 and directed visual inspections of Table 2.

TABLE 2—*The effect of a directed visual inspection at 6600 flight hours.*

Crack Size, cm	Percent Probability, $\times 10^2$	Percent Probability Increment, $\times 10^2$	Average Size, cm	Fraction Missed	Percent Probability after Inspection, $\times 10^2$	Percent Probability after inspection, $\times 10^2$ (cumulative)	Growth Factor	Equivalent Flights at Mean Rate	Crack Size after 1670 Flights, cm
1.25	850					713	1.5	2500	3.5
2.5	280	570	1.88	0.96	547	166	1.8	3000	6.25
5	70	210	3.75	0.70	147	19	2.1	3500	11.75
7.5	15	55	6.25	0.32	18	1	2.4	4000	30
20	3	12	13.75	0.03	1		2.8		
30	1	2	25	0.001			3.0		

NOTES—

Fraction of cracks detected = $\dfrac{850 - 713}{850} = 0.16$.

Probability of detecting a crack = $850 \times 0.16 \times 10^{-2} = 1.4\%$.

128 PROBABILISTIC FRACTURE MECHANICS AND FATIGUE METHODS

TABLE 3—*The effect of a special directed inspection at 6600 flight hours.*

Crack Size, cm	Percent Probability, ×10²	Percent Probability Increment, ×10²	Average Size, cm	Fraction Missed	Percent Probability after Inspection, ×10²	Percent Probability after inspection, ×10² (cumulative)	Growth Factor	Equivalent Flights at Mean Rate	Crack Size after 2860 Flights, cm
0.63	4200	3350	0.94	0.70	2350	2470	0.9	2575	0.75
1.25	850	570	3.75	0.21	120	125	1.5	4300	5.25
2.5	280	210	3.75	0.02	4	5	1.8	5100	10
5	70	55	6.25	0.0012	1	1	2.1	6000	30
7.5	15	12	13.75						
20	3	2	25						
30	1								

Notes—

Fraction of cracks detected = $\dfrac{68 - 904}{6800} = 0.87$.

Probability of detecting a crack = $6800 \times 0.87 \times 10^{-2} = 59\%$.

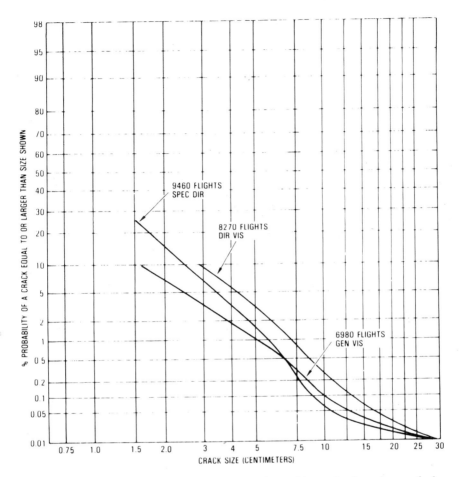

FIG. 6—*Comparison of crack populations at time of second inspection by various methods.*

The procedures outlined previously can be used to estimate the probability of detecting cracks in an individual detail subject to repeat inspections (Col 3). This information and data on the number of independent details subject to this repeat inspection (Col 4) can be used to evaluate the "reasonableness" of the estimated mean time to a practical-to-detect size. Methods of evaluation could include comparisons of the estimated probability of detecting a crack during the accumulated inspections to the number of inspections performed without detecting a crack. Methods of evaluation could also include the number of cracks actually found to the number of cracks that could be expected to be found based on the model.

Ground rules for this type of evaluation may need to be developed and are not explored within the scope of this preliminary study.

TABLE 4—*Accumulated fleet inspections at 16 000 flights on high time aircraft.*

Aircraft Age, 1000 flights	Number of Aircraft in Age Increments	Number of Inspections per Detail	Number of Details Inspected[a]
13	2	3	4
11	3	2	6
0	3	1	6
8	4	1	8
7	8	1	16
6	8	0	0

[a] Assume two details per aircraft.

NOTES—

Inspections: directed visual at intervals of 1670 flights (Table 2).
Inspections threshold: 6700 flights (Fig. 4).

Most important, a mismatch between model and experience should stimulate investigations into basic assumptions such as stress levels and aircraft usage, which greatly influence the estimates of mean crack growth behaviors. In this manner, changes, such as in mean time to a practical-to-detect crack size, would be determined by engineering judgment through reevaluation of model parameters and would be case-peculiar.

Concluding Remarks

It is likely that the data base from which models for fleet inspection rationales can be developed will remain sparse and in some instances incomplete in the forseeable future. There is, however, a great wealth of experience behind current aircraft inspection programs along with a high level of demonstrated safety. Perhaps the best that can be expected of models of the processes at this time is that they agree in general with what has evolved without the formality of a model. In this manner, the models can gain some form of credibility in spite of the minimal data to support assumptions for model components. Once this credibility is gained, an understanding of the model components, their interfaces within the model, and their relationships to inspection experience can guide thinking towards future evolutionary improvements in aircraft inspection programs. The model and discussions of this paper are offered with these thoughts in mind.

F. Kozin[1] *and J. L. Bogdanoff*[2]

Cumulative Damage: Reliability and Maintainability

REFERENCE: Kozin, F. and Bogdanoff, J. L., **"Cumulative Damage: Reliability and Maintainability,"** *Probabilistic Fracture Mechanics and Fatigue Methods: Applications for Structural Design and Maintenance, ASTM STP 798*, J. M. Bloom and J. C. Ekvall, Eds., American Society for Testing and Materials, 1983, pp. 131-146.

ABSTRACT: A new cumulative damage model is described. It is shown how the major sources of variability present in cumulative damage combine in a natural manner to form the general structure of the model and how this structure plays a fundamental role in the assessment of the reliability and maintainability of a component of a structure.

KEY WORDS: cumulative damage, reliability, maintainability, fatigue failure, fatigue (materials), service inspection, replacement policy, stochastic process, mathematical model, probabilistic fracture mechanics, fatigue crack growth, fracture mechanics

Nomenclature

cdf	cumulative distribution function
E	expectation
m	mean of life time
σ^2	variance of life time
$\mu_3(\ )$	third central moment
$\mu_4(\ )$	fourth central moment
x	time in duty cycles (DC)

The purpose of this paper is to describe how B-models [1-7][3] can be used to address the problems of reliability and maintainability in mechanical components subject to fatigue, fatigue crack growth, and wear.

B-models are new. Hence, we will begin by giving a brief description of a simple version of stationary models. Then, we will show how a stationary

---
[1] Professor, Systems, Polytechnic Institute of New York, Farmingdale, N.Y. 11735.
[2] Professor, Aerospace Engineering, Purdue University, West Lafayette, Ind. 47907
[3] The italic numbers in brackets refer to the list of references appended to this paper.

B-model is constructed from life data (synthetic) that pertains to fatigue and fatigue crack growth. The last section is devoted to the study of some of the simpler aspects of problems in reliability and maintainability using the numerical example from the preceding section. In particular, we consider how an inspection schedule, different types of inspection quality, and replacement policy impact on reliability and maintainability of a fleet size, n. Graphical results are given to summarize the numerical findings.

Elements of B-Models

We shall confine attention to just those elements of stationary B-models that will be required. The reader interested in additional details and other aspects of the model is referred to Refs *1* through *7*.

A duty cycle (DC) is a repetitive period of operation in the life of a component during which damage can accumulate. The precise mechanism of the accumulation of this non-negative increment in damage during a DC is not known. A basic assumption is that the increment of damage at the end of a DC depends in a probabilistic manner only on the amount of damage present at the start of the DC, on that DC itself, and is independent of how damage was accumulated up to the start of that DC. This is the familiar Markoff assumption, and we are viewing damage accumulation as an imbedded Markoff process where damage is only considered at the end of each DC.

The time, $x = 0, 1, 2, \ldots$, is measured in number of DC's irrespective of their duration. Damage is indexed by a discrete set of states, $y = 1, 2, \ldots, b$, where State b denotes retirement or failure.

A finite state-discrete time Markoff process can be viewed as a Markoff chain (MC); hence, B-models are framed in terms of MC's. The DC severity is specified in terms of a ($b \times b$) probability transition matrix (ptm), P. In a stationary model, the repetitive DC's have the same severity; this means they have the same ptm, P. A simple form of P is

$$P = \begin{Bmatrix} p_1 & q_1 & 0 & 0 & \cdots & 0 & 0 \\ 0 & P_2 & q_2 & 0 & \cdots & 0 & 0 \\ & & & \cdot & & & \\ & & & \cdot & & & \\ & & & \cdot & & & \\ 0 & 0 & 0 & 0 & \cdots & P_{b-1} & q_{b-1} \\ 0 & 0 & 0 & 0 & \cdots & 0 & 1 \end{Bmatrix} \quad (1)$$

where $p_j + q_j = 1$ and $p_j > 0$, $q_j > 0$. This ($b \times b$) transition matrix has $b - 1$ state dependent transient states, one absorbing state, and only unit-jumps in damage are permitted.

Let the initial state of damage, D_0, a random variable, be specified by the $(1 \times b)$ row vector

$$\vec{p}_0 = \{\pi_1, \ldots, \pi_b\}, \qquad \pi_j = P\{D_0 = j\} \tag{2}$$

$$j = 1, \ldots, b, \qquad \sum_{j=1}^{b} \pi_j = 1$$

We shall assume no component starts in the failed state: thus, $\pi_b = 0$.

The rv D_x denotes the damage at times x; its probability mass function (pmf) is specified by the $(1 \times b)$ vector

$$\vec{p}_x = \{p_x(1), \ldots, p_x(b)\}, \qquad p_x(j) = P\{D_x = j\} \tag{3}$$

$$j = 1, \ldots, b, \qquad \sum_{j=1}^{b} p_x(j) = 1$$

It follows from Markoff chain theory that

$$\vec{p}_x = \vec{p}_0 P^x \tag{4}$$

The cdf, $F_W(x; b)$, of time, W, to failure is given by

$$F_W(x; b) = p_x(b) \tag{5}$$

with reliability and hazard functions defined in the usual manner [1].

Let $F_W(x; j, b)$ denote the cdf of the time, $W_{j,b}$ to reach b given in state j at $x = 0$. Then we can write

$$F_W(x; b) = \sum_{j=1}^{b-1} \pi_j F_W(x; j, b) \tag{6}$$

When $\pi_1 = 1$, Eq 6 becomes

$$F_W(x; b) = F_W(x; 1, b) \tag{7}$$

that is the cdf of $W_{1,b}$ and frequently used. We readily find the moments of $W_{1,b}$ to be [1,6]

$$E W_{1,b} = \sum_{j=1}^{b-1} (1 + r_j) \equiv m_W, \qquad \operatorname{Var} W_{1,b} = \sum_{j=1}^{b-1} r_j(1 + r_j) \equiv \sigma_W^2$$

$$\mu_3(W_{1,b}) = \sum_{j=1}^{b-1} r_j(1+r_j)(1+2r_j), \qquad \mu_4(W_{1,b}) = 3\sigma_W^4 \qquad (8)$$

$$+ \sum_{j=1}^{b-1} r_j(1+r_j)(1+2r_j)(1+3r_j) + \sum_{j=1}^{b-1} r_j^2(1+r_j)$$

where $r_j = p_j/q_j$. Moments of $W_{y,b}$ are found in a similar manner.

Failure may occur before State b is reached. Let the $(1 \times b)$ row vector

$$\vec{\rho} = \{\rho_1, \ldots, \rho_b\}, \qquad \sum_{j=1}^{b} \rho_j = 1 \qquad (9)$$

denote the pmf that defines failure. When failure occurs only in State b, we have $\rho_b = 1$. We assume $\rho_1 = 0$ that means the lowest State 1 cannot be a failed state. Assume for simplicity $\pi_1 = 1$, and let W_1 denote the time to failure. We then find in a manner similar to that used in finding Eq 6 that

$$F_W(x; 1) = \sum_{j=2}^{b} \rho_j F_W(x; 1, j) \qquad (10)$$

where $F_W(x; 1, j)$ denotes the *cdf* of the time to be absorbed in State j given in State 1 at $x = 0$. This formula can be replaced on the right-hand side by $p_x(b)$ if a different form is used for the P given by Eq 1 where the basic process only has unit jumps and $\pi_1 = 1$, [6].

Obviously, we can obtain formulas for time to failure and the corresponding moments when $\vec{p}_0$ with $\pi_1 \neq 1$ and $\vec{\rho}$ with $\vec{\rho}_b \neq 1$ are assumed.

Let there be an inspection at time x_1. Let $\tau_j = $ Prob {damage is detected | damage is in State j}. Assume a component is replaced by a new one if damage is detected. Then it can be shown that

$$P_r^{(1)} = \sum_{j=1}^{b} \tau_j p_{x_1}(j) \qquad (11)$$

is the fraction of components replaced at inspection and the process starts again with initial distribution of damage whose $(1 \times b)$ vector has components

$$P_0^{(1)}(j) = (1 - \tau_j)p_{x_1}(j) + p_r^{(1)}\pi_j, \qquad j = 1, \ldots, b \qquad (12)$$

Further inspections are handled in a similar manner. Other replacement policies can be included as they arise.

This brief presentation summarizes the essential features of the stationary B-model.

The major sources of variability encountered in CD are: (*a*) initial defects due to virgin material, manufacturing processes, inspection, etc.; (b) DC severity that depends on the loading and environment; (c) specification of the damage states for replacement or failure; and (d) the specification of the quality of service inspection and replacement policy. Examination of the preceeding presentation reveals that

(a) is determined by $\vec{p}_0$,
(b) is determined by P,
(c) is determined by $\vec{\rho}$, and
(d) is determined by τ_j plus replacement policy.

Thus, the model assembles into a single structure the major source of variability encountered in CD.

The elements described refer to a stationary B-model. A nonstationary version of B-models is described in Refs 6 and 7.

The power and versatility of B-models is due to the general structure of these models. The parameter sets reveal in explicit form the major sources of variability in every CD process; thus, these parameter sets constitute the power of B-models. Confrontation with data [2,4,6,7] demonstrates the versatility of the models in their ability to describe and analyze life data from a wide variety of sources in fatigue, fatigue crack growth, and wear; further, the parameter estimation problem has in every case proven tractable. Finally, computations are easy to carry out, which is most important in engineering applications.

Let us next apply the results of this section to a numerical example that will permit us not only to illustrate how a model is constructed from data but also to demonstrate the model's use in assessing reliability and evaluating maintainability.

Numerical Example

Let us suppose we have a new component for which we have recorded the times to reach the crack lengths, 0.01, 0.10, 1.00, and 2.50. We assume a precrack phase, that the shortest detectable crack length is 0.01, and that when the crack length is 2.50 the component is no longer useable.

Let us assume sufficient replications were used so we can estimate the mean and variance of the times to reach each of these four crack lengths. Table 1 sum-

TABLE 1—*Mean and variance of time to reach crack length* a.

Crack Length, a	Mean, m	Variance, σ^2
0.01	500	15 630
0.10	1000	71 100
1.00	2000	326 000
2.50	4000	1 306 000

marizes these data (synthetic). We also assume these data indicate that $\pi_1 = 1$ and that the CD process is stationary [6, 7].

Our problem now is to construct a B-model that describes the data of Table 1. The data suggest that $\pi_1 = 1$, so we know $\vec{p}_0$. We want a model with the least number of parameters [8]. This suggests that we use a unit-jump ptm P such as is shown in Eq 1. One way to select the p_j, or equivalently the r_j, to minimize the parameter number, is to assume that there are four blocks of the r_j constant. We find these values of r_j as follows.

Use the first two equations of Eq 8 with $r_j = r_1$, $b = b_1$, and $j = 1, 2, \ldots, b_1 - 1$. Equate these expressions to the means and variance for $a = 0.01$ obtaining

$$500 = (b_1 - 1)(1 + r), \qquad 15\ 630 = (b_1 - 1)r_1(1 + r_1) \qquad (13)$$

An acceptable choice for $b_1 - 1$ is 16 and for $r_1 = 30.25$. Thus, crack length $a = 0.01$ corresponds to Model State 17.

Again use Eq 8 with $r_j = r_1$, $j = 1, \ldots, 16$, $b_1 = 17$, and $r_j = r_2$ for $j = 17, \ldots, b_2$. Equate these expressions to the mean and variance for $a = 0.10$, obtaining

$$1000 = (b_1 - 1)(1 + r_1) + (b_2 - b_1)(1 + r_2)$$
$$71\ 100 = (b_1 - 1)r_1(1 + r_1) + (b_2 - b_1)r_2(1 + r_2) \qquad (14)$$

that have only b_2 and r_2 as unknowns. We find $r_2 = 124$, and $b_2 = 21$. Model State 21 now corresponds to $a = 0.10$.

Containing in this manner, we find

$$r_j = 30.25, \quad = 124, \quad = 249, \quad = 499$$
$$j = 1, \ldots, 16 = 17, \ldots, 20 = 21, \ldots, 24 = 25, \ldots, 28 \qquad (15)$$

with $b_1 = 17$, $b_2 = 21$, $b_3 = 25$, $b = 29$.

It should be pointed up that the b's must be integers. This means we cannot choose the r's to match exactly mean and variance in Table 1 of the times to reach a particular crack length. We chose to match the mean exactly and let the variance deviate from the listed values since variance estimates are less accurate than those for the mean. For example, the variance for $a = 2.50$ in Table 1 is 1 306 000; the model gives 1 314 672, which is close to the required value. We also note that in the first step, $\rho_{b_1} = 1$, in the second step, $\rho_{b_2} = 1$, etc.

We now have P specified. Further, we know that $a = 0.01$ corresponds to Model State 17, $a = 0.10$ corresponds to State 21, $a = 1.00$ corresponds to State 25, and $a = 2.50$ corresponds to State 29. Figure 1 shows a graph of a versus model state. We put a smooth curve through these four points and now

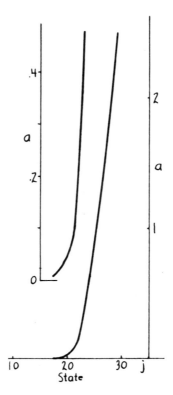

FIG. 1—*Crack length a versus State j.*

can determine for any model state the corresponding crack length, a, but, of course, not conversely, since states are integers.

The data in Table 1 are for four crack lengths. We have chosen to match model mean and variance to these data as accurately as possible using r_1, r_2, r_3, r_4, and b_1, b_2, b_3, b. Sometimes [5], we have data on the times to reach a large number of crack lengths, a. We then could match model mean and variance to the mean and variance (as before) of the times to reach each of these crack lengths. This procedure leads to a large number of parameters. It is shown in Ref 8 that the "best" model uses the least number of parameters consistent with a "good" (ML) description of the data. Thus, it is usually not necessary to match at every crack length for which data are available.

The purpose of this example is to show how the ptm P is obtained when times to reach more than one value of an observable—in this case, crack length—are available. The uniqueness of a model is intimately connected with having an observable as is discussed in Ref 3. It should be noted that we can include a pre-crack phase using B-models; this cannaot be done using other approaches such as da/dn models [5].

Let us now see how we can use our model to discuss reliability and maintainability.

Use of B-Model in Reliability and Maintainability

The reliable performance of a fleet of n units over extended periods of time depends upon a maintenance program. Fundamental to a maintenance program is an inspection schedule that culls out unsatisfactory components and a replacement policy to replace the culled items.

It is easy to state in general terms what must be done to achieve reliable performance of a unit. However, when it becomes necessary to specify the details, the complexity of the task becomes evident as not only are there a vast number of choices available at every step in setting up a program but there are many cost functions that might be used [9,10]. Obviously, in the present paper, we must confine attention to a very simple aspect of the total problem that does, however, indicate the potential use of B-models in problems of this type.

Let us assume that we have a fleet of units of size n in which each unit has one component described by the B-model given in the previous section.

Assume a component has failed when $a = 2.50$ and that it is prudent to remove a component from service when $a = 0.48$ ($j = 23$ from Fig. 1). The appropriate ptm, P, then has its r_j defined by Eq 15. The cdf, F_f, of the time to reach $a = 2.50$ for a new component ($\pi_1 = 1$) with $\rho_{29} = 1$ is obtained from Eqs 4 and 5; this cdf is shown in Fig. 2. The probability $p_r(x)$ that a component needs replacement is obtained from

$$p_r(x) = \sum_{23}^{29} p_x(j) \qquad (16)$$

We also show $p_r(x)$ in Fig. 2. The $p_r(x)$ rises much earlier than F_f, as it should.

Let us assume there are service inspections in the field at times $1000 = x_1$, $2000 = x_2$, $4000 = x_4$, $5000 = x_5$, etc. and factory inspections at $3000 = x_3$, $6000 = x_6$, etc. The interval of 1000 was selected using Fig. 2; we observe from the figure that at $x = 1000\, p_r(x) = 0.11$, which means that on the average 11 percent of the components will require replacement at that time, x_1; and an average of 11 percent replacement is in a reasonable range. We must now use the ideas briefly discussed in connection with Eqs 11 and 12.

Service inspection conducted in the field will detect fewer flaws than will factory inspection. Let the probabilities $\tau_j^{(1)}$ refer to service inspection and the $\tau_j^{(2)}$ refer to factory inspection. We assume the values given in Table 2, where other values of the τ's equal zero. These values clearly indicate that the quality of service inspection is much poorer than is factory inspection.

We also need a replacement policy. Assume that a component culled out in a service inspection is replaced by a used component whose initial distribution of

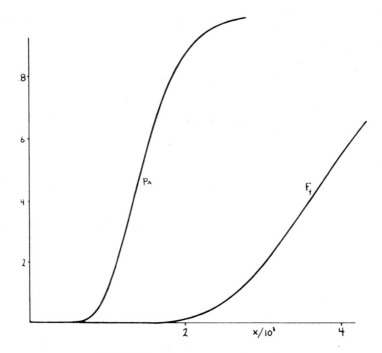

FIG. 2—F_f and p_r versus x; no inspection.

TABLE 2—Probability of detecting a crack of length a.

$\tau_{17}^{(1)} = 1/9,\ \ a = 0.01;\ \ \ \ \ \ \tau_{18}^{(1)} = 2/9,\ \ a = 0.022;\ \ \ \ \ \ \tau_{19}^{(1)} = 3/9,\ \ a = 0.038;$
$\tau_{20}^{(1)} = 4/9,\ \ a = 0.061;\ \ \ \ \tau_{21}^{(1)} = 5/9,\ \ a = 0.10;\ \ \ \ \ \ \ \ \tau_{22}^{(1)} = 6/9,\ \ a = 0.280;$
$\tau_{23}^{(1)} = 7/9,\ \ a = 0.480;\ \ \ \ \tau_{24}^{(1)} = 8/9,\ \ a = 0.700;\ \ \ \ \ \ \tau_{j}^{(1)} = 1,\ \ j \geq 25$
$\tau_{17}^{(2)} = 0.900,\ \ a = 0.01;\ \ \ \tau_{18}^{(2)} = 0.925\ \ \ a = 0.022;\ \ \ \ \tau_{19}^{(2)} = 0.950,\ \ a = 0.038;$
$\tau_{20}^{(2)} = 0.975,\ \ a = 0.061;\ \ \tau_{j}^{(2)} = 1,\ \ j \geq 21,$

damage is specified by the $\pi_j^{(1)}$ and that a component culled out in a factory inspection is replaced by a new component whose initial distribution is given by the $\pi_j^{(2)}$. Table 3 gives the values we shall use. We next need formulas similar to Eqs 11 and 12 in order to complete the description of the problem.

Consider the first service inspection at $x_1 = 1000$. Assume that we replace

TABLE 3—Distribution of damage in replacements.

$\pi_j^{(1)} = j/210,\ \ \ \ j = 1, \ldots, 20;\ \ \ \ \pi_j^{(2)} = 1,\ \ \ \ j = 1$
$\phantom{\pi_j^{(1)}} = 0,\ \ \ \ \ \ \ \ \ \ \ \geq 21\ = 0,\ \ \ \ j \geq 2$

all components detected to have a crack 0.10 or larger. The fraction replaced at x_1 will then be (see Eq 11)

$$p_{r.s}^{(1)} = \sum_{21}^{29} \tau_j^{(1)} p_{x_1}(j) \tag{17}$$

which is the fraction replaced at x_1. The CD process is restarted at x_1 with initial distribution of damage (see Eq 12) given by

$$\begin{aligned} p_0^{(1)}(j) &= p_{x_1}(j) + \pi_j^{(1)} p_{r.s}^{(1)}, & j = 1, \ldots, 20 \\ &= (1 - \tau_j^{(1)}) p_{x_1}(j) + \pi_j^{(1)} p_{r.s}^{(1)}, & j = 21, \ldots, 28 \end{aligned} \tag{18}$$

We note that $p_0^{(1)}(j)$ sums to one. The corresponding formulas for the second inspection are

$$p_{r.s}^{(2)} = \sum_{21}^{29} \tau_j^{(1)} p_{x_2}(j)$$

$$\begin{aligned} p_0^{(2)}(j) &= p_{x_2}(j) + \pi_j^{(1)} p_{r.s}^{(2)}, & j = 1, \ldots, 20 \\ &= (1 - \tau_j^{(1)}) p_{x_2}(j) + \pi_j^{(1)} p_{r.s}^{(2)}, & j = 21, \ldots, 28 \end{aligned} \tag{19}$$

Similar formulas can be written down for all service inspections. It should be observed that above some value of j of the $p_0^{(1)}(j)$ and $p_0^{(2)}(j)$ will be zero.

The first factory inspection occurs at $x_3 = 3000$. The appropriate formulas are

$$p_{r.f}^{(3)} = \sum_{17}^{29} \tau_j^{(2)} p_{x_3}(j)$$

$$\begin{aligned} p_0^{(3)}(j) &= p_{x_3}(j) + \pi_j^{(2)} p_{r.f}^{(3)}, & j = 1, \ldots, 16 \\ &= (1 - \tau_j^{(2)}) p_{x_3}(j) + \pi_j^{(2)} p_{r.f}^{(3)}, & j = 17, \ldots, 28 \end{aligned} \tag{20}$$

In a factory inspection we assume that *all* components detected with a crack are replaced by a new component, which means according to Table 3 that in Eq 20 $\pi_1^{(2)} = 1$ and $\pi_j^{(2)} = 0$ for $j \geq 2$. We note that $p_0^{(3)}(j) = 0$ for $j \geq 21$ and sums to one. Formulas appropriate to subsequent factory inspections follow in an obvious manner from Eq. 20.

We have written down Eqs 17 through 20 by applying Eq 12 without comment. It may assist the reader if we look at some of the features of these equations.

Equation 17 states that all components with a detectable crack 0.10 or larger are removed and the term $\tau_j^{(1)} p_{x_1}(j)$ denotes the probability that damage, D_x, is in State j and damage is detected in State j. The $p_{r,s}^{(1)}$ is the fraction of components culled out at x_1. The $p_0^{(1)}(j)$ denotes the distribution of damage at x_1 *after* inspection and is used to restart the CD process. Thus

$$\vec{p}_x = \vec{p}_0^{(1)} P^{x-x_1}, \qquad x \geq x_1 \tag{21}$$

and this equation is valid until the next inspection. The term $\pi_j^{(1)} p_{r,s}^{(1)}$ in Eq 18 represents that part of the probability of damage being in State j after inspection at x_1 due to the items replaced. The remaining terms on the right of Eq 18 denote that part of the probability of damage being in State j *after* inspection at x_1 due to either not culling out components with Damage State $j \leq 20$ or having missed detecting damage in State 21 or above. Thus, we have the meanings of Eqs 17 and 18 and the terms within. The same meanings apply to Eq 19.

Equation 20 refers to the first factory inspection. Previous comments about this equation plus our comments on Eqs 17 and 18 explain this equation.

We see from Tables 2 and 3 that $p_0^{(1)}(j) = 0$ for $j \geq 25$ and $p_0^{(1)}(j) \neq 0$ for $j \leq 24$. Thus, in service inspection we may have missed some cracks with length 0.48 or greater but less than 1.00, although the probability of doing this is low. This comment refers to every service inspection. The same tables indicate that $p_0^{(3)}(j) = 0$ for $j \geq 21$ and $p_0^{(3)}(j) \neq 0$ for $j \leq 20$, as previously noted. This means that in every factory inspection no crack larger than 0.10 is missed (and replacement is with new components). We now have the physical interpretations of Eqs 17 through 20.

We computed F_f shown in Fig. 2 using the equation

$$F_f(x) = p_x(29), \qquad \vec{p}_x = \vec{p}_0 p^x \tag{22}$$

that is valid for $x \leq x_1$. After the first inspection, we must replace the second by

$$\vec{p}_x = \vec{p}_0^{(1)} P^{x-x_1}, \qquad x_1 \leq x \leq x_2 \tag{23}$$

After subsequent inspections, $\vec{p}_0^{(1)}$ is replaced by the appropriate $(1 \times b)$ vector. Thus, F_f is restarted after every inspection in the normal manner. It is clear that F_f starts from zero after every inspection and then F_f is not the usual cdf that is defined as nondecreasing. However, it still is the probability of failure.

The probability of replacement, $\vec{p}_r(x)$, is obtained using Eq 16 and $\vec{p}_x$ from Eq 23 and similar equations that apply after each inspection. The $p_r(x)$ will not start from zero after every inspection because $\vec{p}_0^{(1)}, \vec{p}_0^{(2)}, \ldots$ have some probability mass in States 23 and 24. Let us display graphically F_f and $p_r(x)$ generated by the preceding technique.

Cases 1 through 4 are shown in Figs. 3 through 6, respectively.

Let Case 1 employ Tables 2 and 3. Let Case 2 be as in Case 1 except for better quality service inspection, where we replace the $\tau_j^{(1)}$ in Table 2 with $\tau_{17}^{(1)} = \frac{1}{5}$,

142 PROBABILISTIC FRACTURE MECHANICS AND FATIGUE METHODS

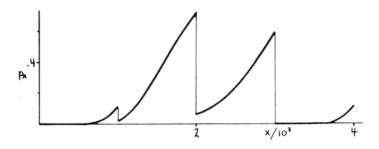

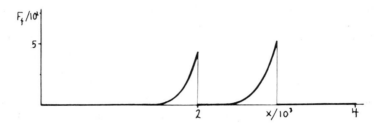

FIG. 3—F_f and p_r versus x; Case 1.

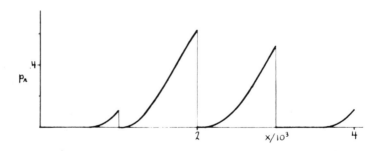

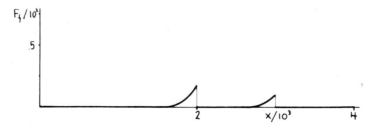

FIG. 4—F_f and p_r versus x; Case 2.

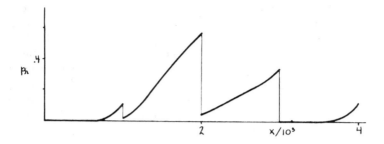

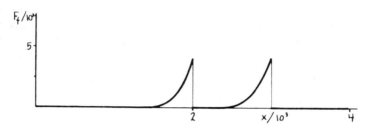

FIG. 5—F_r and p_r versus x; Case 3.

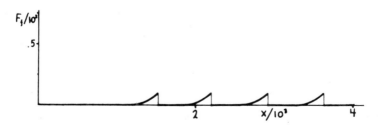

FIG. 6—F_r and p_r versus x; Case 4.

$\tau_{18}{}^{(1)} = \frac{2}{5}$, $\tau_{19}{}^{(1)} = \frac{3}{5}$, $\tau_{20}{}^{(1)} = \frac{4}{5}$, $\tau_j{}^{(1)} = 1$ for $j = 21, \ldots, 29$. Case 3 is the same as Case 1 except we *always* replace culled components with new ones; this means we replace the $\pi_j{}^{(1)}$ of Table 3 with the $\pi_j{}^{(2)}$ of the same table. The interval between inspection times in Cases 1 through 3 is 1000. Case 4 is the same as Case 1 except we call for inspection whenever $F_f = 0.001$ that produces variable inspection times. Numerical results are summarized in Table 4.

We observe from Figs. 3 and 4 that better quality service inspection reduces both $p_r(x)$ and $F_f(x)$; this is also observed in Table 4. While $p_r(x)$ is not much reduced, $F_f(x)$ is substantially reduced in Case 2 over Case 1. Further, $p_r(x)$ is essentially zero after inspection in Case 2 but not for Case 1. The sum $p_{r,s}^{(1)} + p_{r,s}^{(2)} + p_{r,f}^{(3)}$ equals 2.15941 for Case 1 and is 2.49258 for Case 2; this means more components are culled out with better inspection that helps account for the fact that both $p_r(x)$ and $F_f(x)$ are reduced by better quality service inspection.

A comparison of Figs. 3 and 5 reveals that always replacing culled components with new ones also reduces both $p_r(x)$ and $F_f(x)$, as is also seen from Table 4. The sum $p_{r,s}^{(1)} + p_{r,s}^{(2)} + p_{r,f}^{(3)}$, which is the fraction of components replaced during the first cycle of inspection, equals 2.02699; this value is about the same as obtained for Case 1. Inspection of Table 4 and Figs. 4 and 5

TABLE 4—*Reliability and maintenance results.*

x_j	$F_f(x_j)$	$p_r^{(j)}$		$p_r(x_j)$
		Case 1		
1000	5×10^{-6}	0.35395		0.110538
2000	0.004357	0.80851		0.709133
3000	0.005239	0.99695	2.15941	0.585152
4000	13×10^{-6}	0.35544		0.112544
		Case 2		
1000	5×10^{-6}	0.55086		0.110538
2000	0.001740	0.94529		0.623417
3000	0.001026	0.99643	2.49258	0.518387
4000	14×10^{-6}	0.55243		0.112884
		Case 3		
1000	5×10^{-6}	0.35395		0.110538
2000	0.004042	0.68843		0.570638
3000	0.004128	0.98461	2.02699	0.344771
4000	37×10^{-6}	0.36116		0.120007
		Case 4		
1525	0.000995	0.74698		0.578582
2201	0.000994	0.56618		0.380366
2922	0.000992	0.61367	1.9448	0.447111
3625	0.000997	0.60864		0.423066

shows that for the data assumed, the better inspection has a smaller benefit in reliability and maintenance than replacing always with new components.

Figure 6 has variable inspection times chosen so that $F_f(x) \leq 0.001$, everything else being as in Case 1. These times are reasonably close to those used in Fig. 3 for one cycle of inspection, the first being 1525 instead of 1000 but the third is 2922, which is close to 3000. We note the fraction of those replaced after the first cycle of inspection $(p_{r,s}^{(1)} + p_{r,s}^{(2)} + p_{r,f}^{(3)})$ equals 1.9448; this means that fewer components are replaced in the first cycle of inspections while keeping $F_f(x) \leq 0.001$ than in Case 2 but it is not much different from the values in Cases 1 and 3.

It is obvious that we could continue this investigation in a number of other directions such as carrying out more cycles of inspection, using the probability of the first failure in a fleet of size n as our measure of reliability instead as $F_f(x)$, etc. However, enough has been presented for our present purposes.

Discussion and Conclusion

Our objective in this paper has been to demonstrate that B-models can address problems of reliability and maintainability in a meaningful manner.

Reliability and maintainability have to be considered simultaneously since the whole purpose of a maintenance program is to pursue acceptable reliability. This simultaneous consideration ideally requires a general structure of CD, within which not only both aspects naturally fit, but that contains the major sources of variability encountered in CD. B-models have this capability.

We have used a relatively simple numerical example to show that B-models do have this capability.

In the example chosen, we have used the probability of failure as our measure of reliability; obviously other choices can be accounted for within the general structure.

The quality of inspection was easily included and we assumed that service inspection was different from factory inspection. The inspection cycle—service, service, factory—was selected as reasonably representative. Other choices can be made.

We chose a simple replacement policy, namely, service replacement is with used components and factory replacement is with new components. Again, other choices can be made.

Maintenance was assessed using $p_r(x)$—the probability that damage exceeded a specified value (0.48 crack length) and the fraction of components that need replacement at each inspection.

We showed by separate examples the influence on reliability and maintainability of the influence of better service inspection, of a different replacement policy, and of variable inspection times.

The B-model was stationary. Everything we have done could also be accomplished using a nonstationary B-model [7].

The general problem of reliability and maintainability is complex, leading as it naturally must to life cycle cost with all the alternative courses of action available even in simple cases. The cost aspects must be left to future publications. Still we hope by this paper that we have demonstrated that the B-model is a useful tool in addressing this class of problems.

Acknowledgment

The authors gratefully acknowledge the Air Force Office of Scientific Research (AFOSR) support through Contract No. F-49670-78C-0108; I. M. Shimi, control monitor.

References

[1] Bogdanoff, J. L., *Journal of Applied Mechanics*, Vol. 45, No. 2, June 1978, pp. 246-250.
[2] Bogdanoff, J. L. and Krieger, W., *Journal of Applied Mechanics*, Vol. 45, No. 2, June 1978, pp. 251-257.
[3] Bogdanoff, J. L., *Journal of Applied Mechanics*, Vol. 45, No. 4, Dec. 1978, pp. 733-739.
[4] Bogdanoff, J. L. and Kozin, F., *Journal of Applied Mechanics*, Vol. 47, No. 1, March 1980, pp. 40-44.
[5] Kozin, F. and Bogdanoff, J. L., *Engineering Fracture Mechanics*, Vol. 14, 1981, pp. 59-89.
[6] Bogdanoff, J. L. and Kozin, F., *Proceedings*, Annual Reliability and Maintainability Symposium, Jan. 1981, pp. 9-18.
[7] Bogdanoff, J. L. and Kozin, F., *Journal of Applied Mechanics*, Vol. 49, No. 1, March 1982, pp. 37-42.
[8] Akaike, H., *Transactions, Automatic Control*, Institute of Electrical and Electronics Engineers, Inc., Vol. AC-19, No. 6, Dec. 1974, pp. 716-723.
[9] Gnedenko, B. V., Belyayev, Yu, I., and Solovyev, A. D., *Mathematical Methods of Reliability Theory*, Academic Press, New York, 1969.
[10] Nowlan, F. S. and Heap, H. F., "Reliability-Centered Maintenance," Dept. of Defense, Contract No. MDA 903-75-C-0349, Dec. 1978, produced by Dolby Access Press.

C. E. Larson[1] and W. R. Shawver[1]

Method for Determining Probability of Structural Failure from Aircraft Counting Accelerometer Tracking Data

REFERENCE: Larson, C. E. and Shawver, W. R., "**Method for Determining Probability of Structural Failure from Aircraft Counting Accelerometer Tracking Data,**" *Probabilistic Fracture Mechanics and Fatigue Methods: Applications for Structural Design and Maintenance, ASTM STP 798*, J. M. Bloom and J. C. Ekvall, Eds., American Society for Testing and Materials, 1983, pp. 147-160.

ABSTRACT: This study develops a methodology that puts the aircraft's stress at critical or reference locations on a probability basis that includes the variability effects of Mach number, altitude, and gross weight. The stress variation due to these variables is then handled probabilistically and applied to the material's cycles to failure ($S-N$ curve) variation to obtain a fatigue life expended (FLE) probability distribution. This FLE probability distribution may be used to derive FLE values that can be expected for the aircraft instead of the single high conservative value that is now being calculated.

KEY WORDS: fatigue life, probability, aircraft structures, regression analysis, probabilistic fracture mechanics, fatigue (materials), fatigue crack growth, fracture mechanics

Navy aircraft are designed for structural fatigue using various conservative assumptions for protecting structural integrity throughout the useful structural life. In service, a fatigue life tracking system is used for monitoring usage and determining a fatigue life index for individual aircraft. This fatigue life index is used as a measure of the structural life expended. The conservative assumptions involved in establishment of fatigue criteria and fatigue limits are for the effects of variability in material properties, manufacturing processes, chemical environment, and load spectra. The overall effect of the interaction between fatigue design criteria and variability effects is an uncertainty associated with the fatigue life index. Thus, the aircraft may have considerably

[1]Technical project manager and reliability engineering specialist, respectively, Vought Corporation, Dallas, Tex. 75265.

more or less useful life remaining than the calculated fatigue life through the use of risk analysis.

Presently, only the vertical g-load and counting accelerometer (CA) data are used in the algorithms to calculate the fatigue life expended (FLE) value. The FLE is a single-valued quantity arrived at in a deterministic manner and is an excellent activity indicator of aircraft usage. However, the condition of critical structure is dependent upon factors that are stochastic in nature and therefore must be treated probabilistically for a realistic analysis.

The objective of this study was to develop a methodology that puts the FLE on a probability basis that includes the variability effects of Mach number, altitude, gross weight, and material fatigue strength. This developed methodology considers that the stress at a critical or reference location in an airplane is actually a function of several operational variables in addition to the vertical g-load and, through statistical techniques, includes their effect on the fatigue life.

Critical structural locations of Navy aircraft are typically located in the wing. For these cases, such variables as Mach number, altitude, and gross weight are of significance to wing stress levels. Since the counting accelerometer cannot measure these variables, a method is developed that statistically includes their effect in the stress determination at the critical location. The method also recognizes the physical constraints that exist between the variables from a performance standpoint. For example, airplanes are unable to attain high g-loads at arbitrary Mach number, altitude, and gross-weight.

As a background for discussing this methodology development, it is appropriate to first review the current method of determining the FLE.

Fatigue Life Expended Determination, Current Method

The current method of tracking fatigue damage in Navy A-7 aircraft considers the stress at a critical or reference location to be a function of the vertical load factor, n_z.[2] Algorithms that calculate the percent FLE are based on simplified n_z-to-stress relationships in conjunction with Miner's Rule of linear accumulation of fatigue damage. The procedure is illustrated in Fig. 1. Counting accelerometer exceedance data (CA) give the number of times n_z is exceeded at the 5, 6, 7, and 8 g levels. These four data points are interpolated/extrapolated to generate more exceedance data from 3.5 g to 8.5 g in 0.5 g intervals. Exceedance data are then converted to occurrence data by subtracting adjacent points and these compared to cycles to failure data (S-N curve) to determine the fractional damage at each of the load factor levels. These fractions are then summed and doubled to allow for unknown variations that may occur to obtain the FLE.

---

[2]Darnell, J. W., Jr., and Edwards, E. F., "A-7 Wing Service Life Evaluation," Vought Corporation Report 2-53420/OR-5587, 18 Dec. 1970.

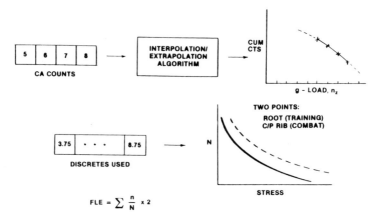

FIG. 1—*FLE determination, current method.*

Fatigue Life Expended Determination, Proposed Methodology

Discrete Case

The proposed methodology to determine the FLE probability level incorporates the effects of Mach number, altitude, and gross weight variation. Using regression analysis with these variables and the g-load, the values for the unknown coefficients are estimated. The general expression for the assumed mathematical relationship is

$$S = A_0 + A_1(g) + A_2(\text{MN}) + A_3(\text{ALT}) + A_4(\text{WT}) + \epsilon \qquad (1)$$

where

- S = stress at reference location, MPa,
- g = g-load, g,
- MN = Mach number,
- ALT = altitude, m,
- WT = gross weight, kg,
- ϵ = random error due to random fluctuation of S not expressed by the other predictor variables, assumed to be independent, normally distributed with zero mean and constant variance, and
- $A_0, A_1, \ldots A_4$ = unknown coefficients that are solved for through regression.

Thus, for a given g-load and simultaneous values of MN, ALT, and WT, the stress at a reference location can be estimated. However, if the values of

MN, ALT, and WT are not measurable along with the g-load but it is desirable to include their effect in the analysis, then the following methodology is proposed.

If the mean values of MN, ALT, and WT are substituted into Eq 1 then a new equation for stress as a function only of g-load is obtained

$$S = A_0' + A_1(g) + \epsilon' \qquad (2)$$

where A_0' denotes a new constant containing the MN, ALT, and WT mean values; and ϵ' denotes a new random error that now includes the effects of the MN, ALT, and WT as they deviate from their mean values.

The variance about the regression line of stress on g-load may now be considered to be

$$\sigma_S^2 = A_2^2 \sigma_{MN}^2 + A_3^2 \sigma_{ALT}^2 + A_4^2 \sigma_{WT}^2 + (SEE)^2 \qquad (3)$$

where

$\sigma^2 =$ variances about the subscripted variables, and
SEE $=$ standard error of estimate.

These values are obtained from the regression analysis of Eq 1.

The graph of Eq 2 with the effects of MN, ALT, and WT included within the confidence bands is illustrated in Fig. 2.

The same results can be obtained by regressing stress on g-load only, as per Eq 2. In fact, it is necessary to compute the regression in this manner if the

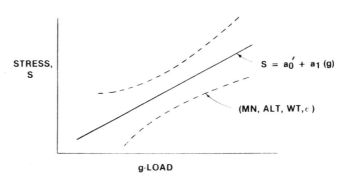

FIG. 2—*FLE determination, regression analysis.*

MN, ALT, and WT can not be measured simultaneously with the counting accelerometer. However, the proposed methodology deals not with the development of the regression equation but with the application of it. The current methodology uses the regression equation to estimate a single stress value for a given g-load and then applies this single stress value in subsequent calculations. This method proposes to use the variance of the regression analysis of Eq 2 to define the stress probability distribution for a given g-load and, by so doing, to include the effects of the MN, ALT, and WT knowing that their effects are embedded within the stress probability distribution. This approach also takes care of the physical constraints between the variables, because the flight data for the regression analysis contains only data for which the variables are compatible.

The proposed methodology is illustrated in Figs. 3 and 4 and sequentially described in the following four steps.

Step 1—The actual counting accelerometer settings as shown in the squares of Fig. 3 are used in the interpolation/extrapolation algorithm for additional values. As illustrated for a given g-load interval, Δg, the number of occurrences, n_i, for that interval is obtained by subtracting the two cumulative number of exceedances counts. With this n_i is a corresponding g_i, the midvalue of the interval, as shown in the right-hand-side figure. In this manner, the g-load range is divided into 0.5 g intervals, and a discrete set of g_i, n_i values is obtained.

Step 2—Assuming that the variance of S, σ_S^2, is constant for all values of g and that S is normally distributed, then for each g_i there is a range of S-values defined by the frequency distribution, $F(S)$, that may be estimated by

$$F(S) = S_i + Z(\text{SEE}) \tag{4}$$

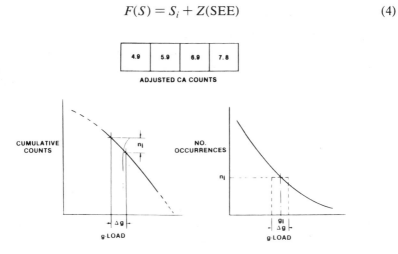

FIG. 3—*FLE determination, counting accelerometer data handling.*

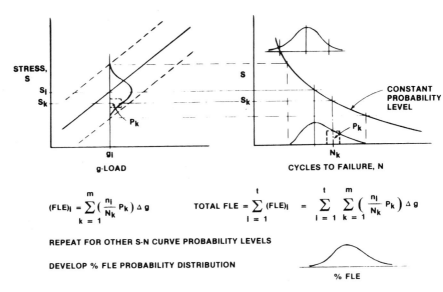

FIG. 4—*FLE determination, probability method.*

where

$F(S)$ = frequency distribution of S,
S_i = value of stress from the regression,
Z = standard normal variable, and
SEE = standard error of estimate from the regression of stress on g-load.

This is illustrated in the left-hand-side of Fig. 4. Divide the range of S shown with its probability density function into intervals and denote the mid-value of each interval S_k. The probability of its occurrence is P_k. As illustrated, apply this S_k to the S-N curve and obtain its associated cycles to failure value, N_k. The N_k will also have the same P_k value.

Step 3—Apply Miner's Rule as shown in the equation

$$(\text{FLE})_i = \sum_{k=1}^{m} \left(\frac{n_i}{N_k} P_k\right) \Delta g_i \tag{5}$$

where $(\text{FLE})_i$ is for the discrete pair of g_i, n_i obtained in Step 1. Notice the summation is over the stress range from Step 2, for the constant, g_i. The total FLE is then calculated by summing the $(\text{FLE})_i$ over the g_i

$$\text{TOTAL FLE} = \sum_{i=1}^{t} (\text{FLE})_i = \sum_{i=1}^{t} \sum_{k=1}^{m} \left(\frac{n_i}{N_k} P_k\right) \Delta g_i \tag{6}$$

Step 4—Repeat Steps 2 and 3 for other S-N curve constant probability levels. The S-N curve used is at a constant probability level that is determined in the material strength data analysis. The chosen probability levels used in the S-N curve give those probability values to the TOTAL FLE calculated. Thus, by selecting other S-N curve constant probability levels for the N_k values and repeating the calculations, a probability distribution for the FLE is obtained as shown on the lower right-hand-side of Fig. 4.

Continuous Case

A continuous case or closed-form solution can be used in the methodology for determining the TOTAL FLE. The solution is illustrated in Fig. 5. Equations for the following continuous functions in Fig. 5 need to be determined, namely,

    a. number of occurrences versus g-load, $n = f_1(g)$ (*left*);
    b. stress probability distribution, $f_2(S)$, about its regression line (*middle*); and
    c. cycles to failure in terms of stress, $N = f_3(S)$ (*right*).

As can be seen in Fig. 5, by allowing the number of intervals in the summations to approach infinity, we obtain for the limits the double integral shown where the limits for the first integral are over the range of g-loads and the limits for the second integral are over the stress distribution about its regression line.

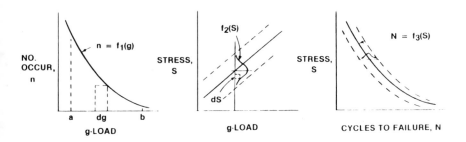

FIG. 5—*Closed form solution.*

154   PROBABILISTIC FRACTURE MECHANICS AND FATIGUE METHODS

Numerical Example

For a numerical example to illustrate the proposed methodology, an A-7E Navy aircraft with 1698 flight hours accrued, which had been published in NADC's quarterly report[3], was used. The four steps of the methodology will be followed in the example.

Step 1—The cumulative number of counts that the g-levels were exceeded are shown in Fig. 6 inside the boxes. Also shown are the g-level setting of the recorders that are used in the analysis. The four readings are interpolated/extrapolated using the first column g_i values in Fig. 6 to produce the pair of g_i, n_i shown in the last two columns. As illustrated in the figure, the g_i represents the mid-value of each interval and n_i is the number of occurrences for the interval. Calculations for a typical pair of values, $g_i = 5.15$ and $n_i = 719$, as illustrated, will be carried through in the next step.

Step 2—The critical location for the A-7E aircraft was taken to be at Wing Station 53.7. At this station the regression of stress on g-load was assumed to be

$$\text{stress} = 27.57(g)$$

Confidence interval for the equation was calculated from Searle[4]

$$S = \bar{S} + A_1(g - \bar{g}) \pm t_{N-2,\,1/2\alpha}\,\hat{\sigma}_s \left[\frac{1}{N} + \frac{(g - \bar{g})^2}{\sum_{i=1}^{N}(g_i - \bar{g})^2} \right] \quad (7)$$

where

S = stress, MPa,
g = g-load, g,
$\bar{S}$ = mean value of stress, MPa,
$\bar{g}$ = mean value of g-load, g,
A_1 = Regression coefficient,
$\hat{\sigma}_S$ = square root of the predicted error variance of the stress,
$t_{N-2,\,1/2\alpha}$ = $100\,(1 - \alpha/2)$ percent point from a student t-distribution with N-2 degrees of freedom, and
N = number of data points.

In our example, the predicted error variance is expressed by the standard error of estimate from the regression, namely,

$$\hat{\sigma}_S^2 = (\text{SEE})^2$$

---

[3]Virga, R. R., "Aircraft Structural Appraisal of Fatigue Effects (SAFE) Program," NADC-13920-1, Naval Air Development Center, 15 April 1980.
[4]Searle, S. R., *Linear Models*, Wiley, New York, June 1971, p. 108.

A-7E,   1,698 Flight Hours

4.9	5.9	6.9	7.8
1,262	204	18	2

ADJUSTED CA COUNTS

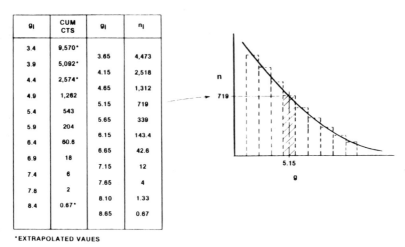

g_i	CUM CTS	g_i	n_i
3.4	9,570*		
		3.65	4,473
3.9	5,092*		
		4.15	2,518
4.4	2,574*		
		4.65	1,312
4.9	1,262		
		5.15	719
5.4	543		
		5.65	339
5.9	204		
		6.15	143.4
6.4	60.6		
		6.65	42.6
6.9	18		
		7.15	12
7.4	6		
		7.65	4
7.8	2		
		8.10	1.33
8.4	0.67*		
		8.65	0.67

*EXTRAPOLATED VAUES

FIG. 6—*Numerical example, data handling.*

Earlier analysis of 54 000 data points was made on the A-7D and the SEE was calculated to be 5.96. This value was used for our example, since no data for the A-7E were available.

The frequency distribution, $F(S)$, about the predicted stress, S_i, for a given g_i is estimated by

$$F(S) = 27.57(g_i) + Z(5.96)$$

The graph of the regression equation with ± 3 (SEE) bands is given in Fig. 7. The frequency distribution was divided into eight equal stress intervals and the P_k values calculated from a normal distribution. Sum of the P_k values must equal 1. The values for the 4.9 to 5.4 g interval are tabulated in Fig. 8. The N_k values for the given S_k were taken from the assumed S-N curves shown in Fig. 9, for a constant probability level at the lower 10 percent, designated ①. This is a basic material's property curve that would need to be determined for the methodology.

Step 3—Miner's Rule is next applied as illustrated in Fig. 8. This sum, 0.0152, is for the constant $g_i = 5.15$. The TOTAL FLE is then calculated for a constant probability level by summing the $(FLE)_i$ over the g_i's, as illustrated in Fig. 10 in the first two columns.

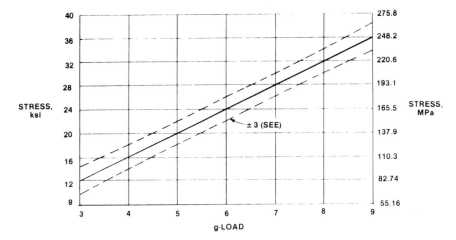

FIG. 7—*Regression of stress on g-load at WS 53.7 for A-7E aircraft.*

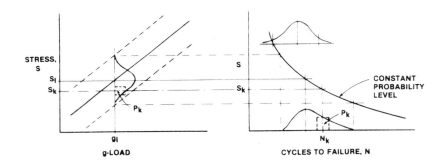

FIG. 8—*Numerical example, interval calculations.*

FLE FOR 4.9 - 5.4 g INTERVAL — AT LOWER 10% PROBABILITY					
g_l	n_l	P_k	S_k	N_k	$\dfrac{n_l}{N_k} P_k$
5.15	719	0.01	126.3	170,000	0
		0.06	130.8	100,000	0.0004
		0.16	135.3	68,000	0.0017
		0.27	139.8	53,000	0.0037
		0.27	144.3	43,000	0.0045
		0.16	148.8	36,000	0.0032
		0.06	153.2	31,000	0.0014
		0.01	157.7	27,000	0.0003
		1.00			0.0152

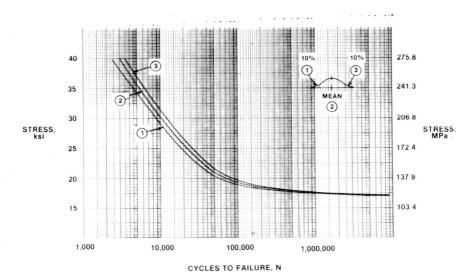

FIG. 9—*Stress versus cycles to failure, typical material property curve (assumed probability distribution).*

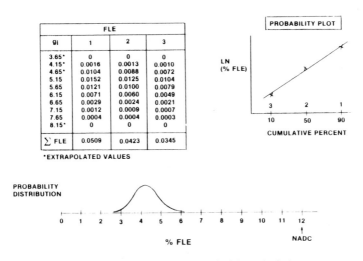

FIG. 10—*Numerical example, probability calculations.*

Step 4—New sets of N_k values based on 50 percent and upper 90 percent constant probability levels, designated ② and ③, respectively, are shown in Fig. 9. Miner's Rule is applied to these as in Step 3, and the results are shown in Fig. 10, in the last two columns. Natural logarithms were taken on these three calculated FLE values and plotted on probability paper making a cumulative frequency plot as seen in Fig. 11. A straight line represents these three points

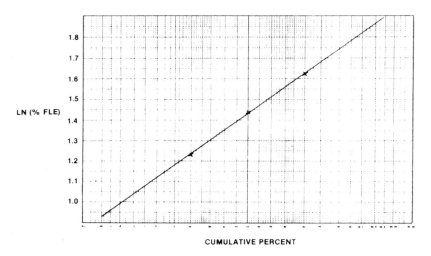

FIG. 11—*FLE log normal cumulative frequency plot.*

reasonably well indicating the FLE probability distribution may be represented by the logarithmic normal. This defines the probability density function as illustrated in the bottom of Fig. 10. Because of the small range of values, the skewness of the logarithmic normal distribution is not very apparent. The range shown is ± 3 (standard deviation), that is, $\pm 3\,\sigma$. The NADC quarterly report showed the FLE for the aircraft to be 12 percent. This implied a calculation of 6 percent that is to be expected for their method uses a conservative S-N curve with values in the lower $3\,\sigma$ neighborhood.

Results and Discussion

1. As shown in the numerical example, an FLE probability distribution can be obtained that includes the effects of MN, ALT, and WT. These effects, however, are embedded within the error variance of the stress on g-load regression and are dealt with jointly. The separate effects cannot be obtained without measured data that are not available in our case.

It should be mentioned here that earlier regression analyses of stress on the independent variables, g-load, MN, ALT, and WT, had obtained correlation coefficients above 0.90 and had established that these independent variables are statistically significant. Thus, we are justified in claiming their effects are in the error variance and are simply presenting a way of determining their joint effect.

2. The FLE probability distribution provides much more information for risk assessment than the single value that has been doubled to cover effects from the unknown variables. When the single conservative number shown in

Fig. 10 of 12 percent begins approaching 100 percent, the FLE probability distribution, obtained by the ratio 100/12, will range from 25 to 50 percent. This distribution then gives a measure for the conservatism that is built into the single-value method and no doubt would aid the analyst in selecting a course of action. Without this additional information, the analyst must decide either to suspend the aircraft's operation, which may be premature, or, knowing the FLE is a conservative value, allow it to continue flying with an unknown risk of failure.

3. The analysis may be formulated in closed form. However, the equations may be difficult to integrate. The discrete analysis is easy to do by hand calculator or computer and is believed to be a satisfactory approximation considering the assumptions used in the analysis.

4. The material S-N curve with its probability distribution is critical to the analysis. We did not have one available and had to assume a distribution for the example. Before applying the methodology, this information would need to be established.

5. In the FLE table of Fig. 10, the first three and the last g-load values are extrapolated data points. These values represent approximately 24 percent of the total FLE that may or may not be acceptable to the analyst. Feedback of this information can be used to help set the levels of the counting accelerometer registers on the aircraft fleet.

Conclusions

The methodology provides more visibility in risk assessment for the following reasons.

  a. Gives FLE probability distribution instead of a single value.

  b. Includes variation effects due to: Mach number, altitude, gross weight, and material properties.

  c. Gives a more realistic value to the FLE than by doubling the FLE calculations to cover effects from unknown variables.

Acknowledgments

This methodology was developed for a program funded by the Naval Air Systems Command under Contract N00019-80-C-0299[5]. The program was monitored and directed by A. Somoroff and M. Dubberly of AIR 5302, and was conducted by the Vought Structural Life Assurance and Reliability Engineering Groups under the direction of C. E. Larson and J. T. Stracener, respectively.

---

[5]Etchart, M. P., Merkord, D. L., and Shawver, W. R., "Structural Risk Methodology Development," Vought Corporation Report 2-51220/2R-53034, 3 Feb. 1982.

DISCUSSION

H. S. Reemsnyder[1] (written discussion)—Mr. Shawver must be complimented on his clear presentation and effective visual aids. As I understand the presentation, the frequency distribution of stress for a given acceleration in g's was constructed from the confidence interval about the linear regression of σ on g. The 100γ percent *confidence interval* for the regression line may be defined as follows. A sample of n g-σ pairs, is selected and its regression line and associated γ percent confidence interval are constructed. The width of the confidence interval is so determined that, in many repetitions of this whole procedure (selecting a sample, and constructing both the regression line and its γ percent confidence interval), 100γ times out of 100, the confidence interval will include the regression line of the entire population.

On the other hand, assuming that the variance of σ is constant for all values of g and that σ is normally distributed, the frequency distribution $F(\sigma)$ of σ may be estimated by

$$F(\sigma) = \sigma' + z \cdot s_{\sigma/g}$$

where σ, z, and $s_{\sigma/g}$ are, respectively, the value of σ from the regression, the standardized normal variable, and the standard error of estimate from the regression of σ on g. If a percentile $F(\sigma)$ for the population of all σ is to be estimated, then the lower tolerance limit should be used to introduce the uncertainty due to sample size, etc.

In my opinion, it is incorrect to estimate the frequency distribution of the dependent variable from the confidence interval about the regression of that variable on another, independent variable. Instead, the frequency distribution should be estimated in a manner similar to that just described.

C. E. Larson and W. R. Shawver (authors' closure)—Mr. Reemsnyder's comments are insightful and well-founded. They have been incorporated in Step 2 of the Proposed Methodology.

---

[1] Research Department, Bethlehem Steel Corporation, Bethlehem, Pa. 18016.

C. E. Bronn[1]

Analysis of Structural Failure Probability Under Spectrum Loading Conditions

REFERENCE: Bronn, C. E., "**Analysis of Structural Failure Probability Under Spectrum Loading Conditions,**" *Probabilistic Fracture Mechanics and Fatigue Methods: Applications for Structural Design and Maintenance, ASTM STP 798*, J. M. Bloom and J. C. Ekvall, Eds., American Society for Testing and Materials, 1983, pp. 161-183.

ABSTRACT: A method is presented for computing the probability of failure of structural members exposed to spectrum loading conditions and subjected to periodic inspections.
The output is failure probability as a function of service exposure time. The effects of perturbations in the severity of service exposure and variations in inspection effectiveness and inspection period length are demonstrated.

KEY WORDS: failure probability, spectrum loading, crack growth, initial equivalent flaw distribution, periodic inspection effects, fatigue (materials), fracture mechanics, probabilistic fracture mechanics

During the last decade there has been a steadily increasing trend toward complementing deterministic aircraft design criteria with probabilistic considerations, initially in the areas of systems design [1,2].[2]

Much earlier, dating back to the middle and late thirties, fatigue phenomena began to appear in the area of airplane structures and were recognized by authorities as essentially probabilistic in nature by the early fifties.

The classical Palmgren-Miner concept of cumulative damage represented a first attempt at evaluating fatigue endurance. However, due to the concept's deterministic nature that restricted its application to estimations of mean values for fatigue endurance, it required statistical tack-ons to adequately describe probabilistic phenomena. This was generally achieved by means of scatter factors derived from test experience [3-5]. It is now being supplanted by analysis methods based on fracture mechanic's concepts and principles [6].

---
[1]Deceased; formerly, staff engineer, Lockheed Georgia Company, Marietta, Ga. 30063.
[2]The italic numbers in brackets refer to the list of references appended to this paper.

The approach presented in this paper is based on fracture mechanics. In it, the crack growth and fracture characteristics are considered to be deterministic functions of crack size and spectrum loading exposure. The probabilistic aspects are represented by equivalent initial flaw distributions and load exceedance probabilities. It was developed for application to airplane structures exposed to loading conditions conducive to fatigue-fracture phenomena.

Definitions

1. A structural member is defined as a single, detachable part of the structure exposed to a substantially uniform level of gross area stress.
2. A failure is defined as the complete severance of the member.

Basic Assumptions

The analysis rests on the following basic assumptions:

1. A failure is caused by one crack and one crack only, namely, the largest.
2. In any structural member containing multiple cracks, the rate of the crack growth will be higher for larger than for smaller cracks. Consequently, the largest crack at any time will remain the largest throughout the life of the member, unless it is removed by repair action.
3. Structural members will, in general, possess a finite number of local stress raisers such as fastener holes and cross-sectional shape changes. Each of these stress raiser sites contains a large population of initial flaws. One of these flaws is the largest in the sense that its size, shape, orientation, and location in the local stress field permits it to grow into the largest observable crack under operating stress conditions.
4. Fracture mechanics methodology can be applied to reduce observed cracks with varying characteristics (depth, aspect ratio, location) to equivalent quarter circle corner cracks. The distribution of quarter circle crack radii obtained in the preceding manner is called the distribution of equivalent crack sizes.

Description of Computation Input Data

The methodology is illustrated by example data applicable to the wing surface of a military transport aircraft.

Maximum Stress Exceedance Data

Figure 1 shows the type of data used in the analysis. It is presented as log [exceedance probability] [log $F_E(S)$] versus levels of maximum stress, S. The

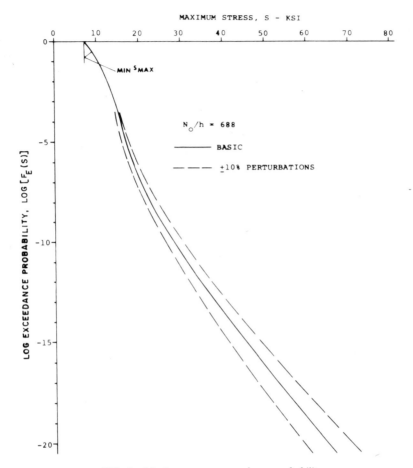

FIG. 1—*Maximum stress exceedance probability.*

curve was generated by dividing the expected exceedance frequencies at various maximum stress levels by the expected frequency of exceeding the minimum maximum stress, $_{min}S_{max}$. The stress exceedance event frequency N_0/h, is the number of exceedances of maximum stress equal to zero per aircraft service life, (N_0) divided by the life time in hours (h). The stress level at probability 1.0 (log (probability) = 0) is the minimum maximum stress level for the mission profile, defined as the lowest 1.0 g maximum stress for any mission within the profile. It and all other maximum stress levels will be exceeded with a probability of one. The two dashed lines bracketing the solid line (basic) represent ±10 percent perturbations to the difference ($S - {_{min}S_{max}}$) used to investigate the effect of load spectrum perturbations on the failure probability.

Residual Strength Data

Figure 2 shows the residual strength data used in the example problem. Shown is a graph of the critical stress, S_{crit}, at each crack length. In generating the data, only the net crack length was considered, yielding a monotonically decreasing strength with increasing crack length.

Crack Growth Data

Figure 3 and Table 1 show, in graphical and tabular forms, an example of the crack growth data used in the analysis. The data were generated under the assumption of initial simultaneous double-edged crack growth until ligament break-through, followed by a single-edged crack growth into the member from a crack length equal to twice the ligament width.

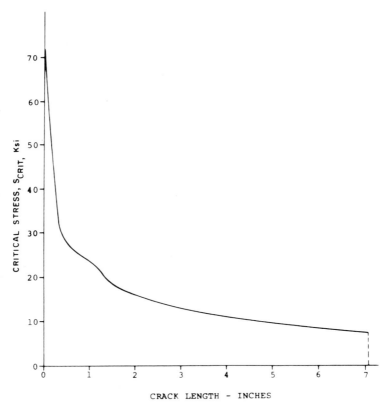

FIG. 2—*Residual strength (critical stress).*

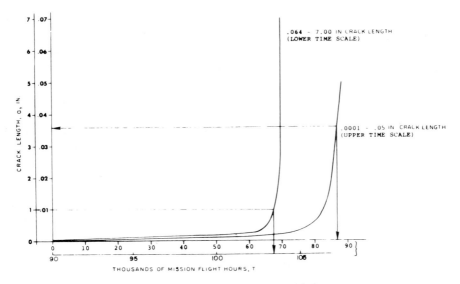

FIG. 3—*Crack growth curve, parametric* da/dN.

TABLE 1—*Crack growth.*

a, in.	T, h	a, in.	T, h	a, in.	T, h
0.0001	0	0.0319	86 418	0.4500	103 090
0.0010	50 000	0.0383	87 185	0.5000	103 138
0.0012	56 965	0.0460	88 013	0.6000	103 224
0.00144	62 452	0.0500	88 438	0.7000	103 301
0.00173	66 786	0.0600	89 532	0.7500	103 336
0.00207	70 155	0.0720	90 890	0.8000	103 371
0.00249	72 831	0.0864	92 463	0.9000	103 438
0.00299	75 006	0.1000	93 775	1.0000	103 498
0.00358	76 795	0.1037	94 132	1.2500	103 611
0.00430	78 285	0.1244	95 841	1.5000	103 692
0.00516	79 533	0.1493	97 615	1.7500	103 752
0.00619	80 572	0.1500	97 661	2.0000	103 798
0.00743	81 443	0.1800	99 476	2.2500	103 832
0.00891	82 200	0.2000	100 395	2.5000	103 859
0.01070	82 877	0.2160	101 130	2.7500	103 881
0.0128	83 496	0.2500	101 929	3.0000	103 899
0.0154	84 083	0.2590	102 140	4.0000	103 914
0.0185	84 665	0.3000	102 663	5.0000	103 922
0.0222	85 255	0.3300	102 902	6.0000	103 926
0.0266	85 851	0.3630	103 035	7.052	103 930

Initial Equivalent Flaw Distribution

Data Source—The data used in this analysis were derived from the results of the tear-down inspection performed on a service aircraft. Upon disassembly, the several flaw sites (fastener holes) were subjected to thorough microscopic investigations. Existing flaws or cracks were measured, recorded, and categorized according to the classification scheme shown in Fig. 4. The parameters, a and w, for individual flaws were next input in an analytical crack growth program utilizing a two-layer stress spectrum. The program was exercised until an 0.3-in. long surface crack was reached and the required number of cycles was recorded. Back-calculating the same number of cycles from an 0.3-in.-long surface crack grown with the same spectrum from a quarter circle corner flaw established the size of the equivalent quarter circle corner crack reflecting the structural status of the site at the time of the tear down.

To determine the initial equivalent quarter circle flaw size, a second regression was performed, using the exposure status (mission profile hours) of the

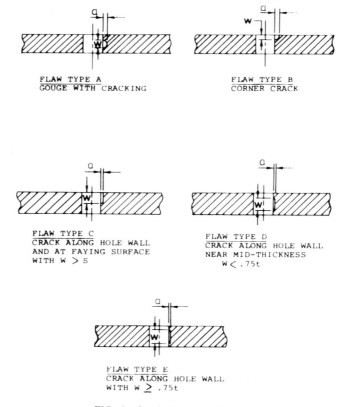

FIG. 4—*Crack shape classification.*

member at the time of tear down inspection along with the mission profile and appropriate crack growth data. The output of these activities were:

(a) Total number of flaw sites in the member, N.
(b) Number of flaw sites inspected, N_I.
(c) Number and sizes of equivalent quarter circle cracks at time of teardown, N_F, X_i.
(d) Number and sizes of initial equivalent quarter circle flaws, N_F, a_i.

Statistical Analyses—The procedure just outlined yielded in most cases one, sometimes two, and very rarely three or more cracks per damaged site. In accordance with preceding item (c), the collection of flaws found could therefore be regarded as representing an extreme (largest) distribution of flaw sizes for the inspected sites.

The objective of the statistical analysis was to obtain the probability distribution function for the largest flaw in a member with N sites from the characteristics of the flaw collection that was found by inspection of N_I sites.

Denoting the number of discovered flaws by N_F, and following Ref 7, the flaws were ordered in the sense of increasing size, a_i, and assigned rank order numbers R_i given by

$$R_i = N_I - N_F + i \qquad (i = 1, 2 \cdots N_F) \tag{1}$$

The probability distribution function of values (extreme values) for the flaws could then be evaluated as

$$F_{N_I}(a_i) = P(a \leq a_i) = R_i/(N_I + 1) \tag{2}$$

And for the member with N sites

$$F_N(a_i) = [R_i/(N_I + 1)]^{(N/N_I)} \tag{3}$$

Analytical Representation—The analytical representation had to satisfy two requirements:

(a) It had to afford a faithful rendering of the observed data over the range of observations.
(b) It had to provide the best possible representation of data trends for extrapolation into the range of large flaw sizes.

The following transformations were found to be quite successful in satisfying these two requirements

$$U_i = \ell n\{\ell n[1/(1 - F_N(a_i))]\} \tag{4}$$

$$V_i = \ell n(1/a_i) \tag{5}$$

Plots of these transformed variables are shown in Fig. 5. It is evident that the data ranges can be represented by two or three linear segments of the form

$$U = A_k^0 + B_k^0 \cdot V \qquad (k = 1, 2, 3) \tag{6}$$

with different coefficient sets A_k^0 and B_k^0 for each segment. From the standpoint of trend extrapolation, it also appears that the upper data trend (low V-values) is well represented by the linear form of Eq 6. It is noted that the zero-time regression appears to reduce the data scatter.

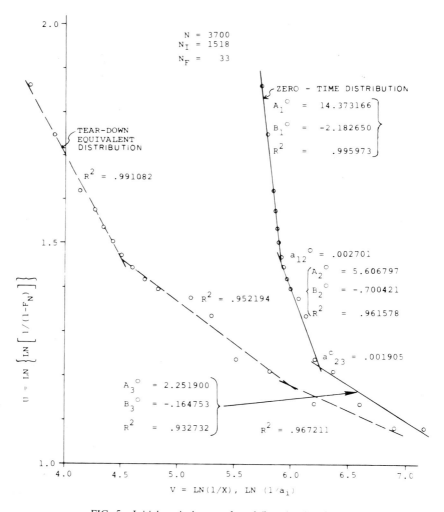

FIG. 5—*Initial equivalent crack and flaw size distributions.*

Inversion of the transformations in Eqs 4 and 5 yields

$$F_N(a_i) = P(a \le a_i) = 1 - e^{-e^{[A_k^0 + B_k^0 \cdot (\ln(1/a_i))]}} \qquad (k = 1, 2, 3) \qquad (7)$$

that will be recognized as a form of the Weibull distribution with segmentwise changing constants. (Actually, Eq 7 is an extreme (smallest) value distribution in the inverted variate $\ln(1/a_i)$, and consequently an extreme (largest) distribution in the variate a_i.)

For the purpose of failure probability computations, a more useful quantity is the flaw size exceedance probability $F_{EN}(a_i)$:

$$F_{EN}(a_i) = 1 - F_N(a_i) = e^{-e^{[A_k^0 + B_k^0 \cdot (\ln(1/a_i))]}} \qquad (k = 1, 2, 3) \qquad (8)$$

A comparison of the actual and analytically represented data is shown in Fig. 6. (For convenience, log (probability) is shown.)

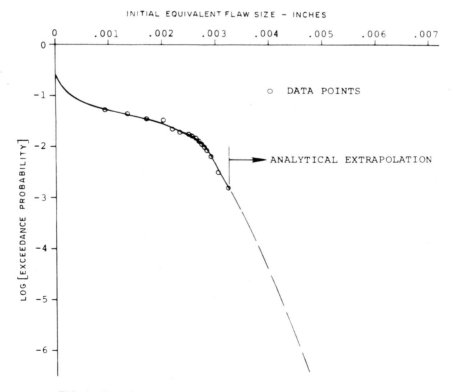

FIG. 6—*Exceedance probability distribution for initial equivalent flaw sizes.*

Crack Detection and Repair Probability Data

Crack Detection—Reference *8* contains results from an Air Force program to experimentally determine the reliability of several nondestructive inspection methods applied in a variety of inspection situations. The results, expressed in terms of detection probability, P_D, for varying crack lengths, X, could in general be expressed by the relationship

$$P_D(X) = 10^{-[A \cdot (1/X)^{B-1}]} \qquad (9)$$

with Coefficients A and B determined by a linear regression of transformed variables. The reported data did not extend appreciably beyond crack lengths of 1.0 in. However, for the present purpose, it is necessary to estimate the detection probability for much larger cracks, of the order of 7 in. For the purpose of demonstration, Eq 9 was used for such extrapolations, with Coefficients A and B taken from Ref *8*. Again for strictly illustrative purposes, the data used here are the mean performances taken from Figs. 10-1 for eddy current surface scan (ECSS) inspection and from Fig. 10-3 for ultrasonic surface wave (USSW) inspection in Ref *8*.

Repair Probability—No data were available on the incidence of missed repair opportunities. Consequently, it will be assumed that whenever a crack is discovered, it will also be repaired, that is, repair probability

$$P_R = 1.0 \qquad (10)$$

The probability of a crack, X, being detected and repaired at a maintenance event is consequently

$$P_{\text{fix}}(X) = (1.0) \cdot P_D(X) \qquad (11)$$

As will be shown in a later section, the important parameter from a failure probability standpoint is the probability $P_s(X)$, that the crack, X, will escape detection and repair at a maintenance event.

$$P_s(X) = 1 - P_{\text{fix}}(X) = 1 - P_D(X) \qquad (12)$$

Figures 7 and 8 show the mean detection probabilities taken from Ref *8*, with the probability of the crack not being detected and repaired shown in Fig. 9 using a log scale because of the wide range of the variable.

Description of Computation Steps

Failure Probability Computations

Analysis—The analysis follows closely the concepts presented by Chilver [*9*].

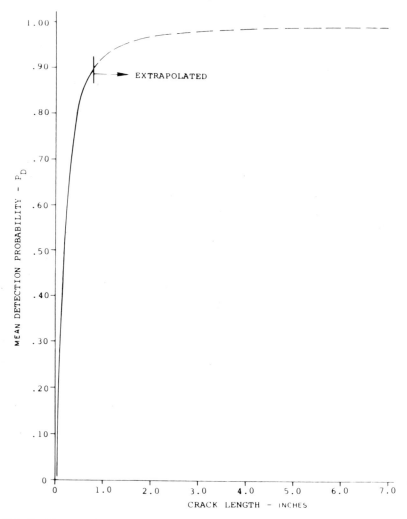

FIG. 7—*Mean crack detection probability for eddy current surface scan inspection (Ref 8, Fig. 10-1).*

The occurrence probability, $\Delta P(x, \Delta X)$, at time T of a crack in the size range X, $X + \Delta X$ is identical with the occurrence probability of an equivalent initial flaw in the size range such that it would grow into a crack in the size range X, $X + \Delta X$ in T hours. This occurrence probability is

$$\Delta P(X, \Delta X) = F_{EN}\{a_i[(X + \Delta X), T]\} - F_{EN}[a_i(X, T)] \qquad (13)$$

(Note: $X + \Delta X < X$)

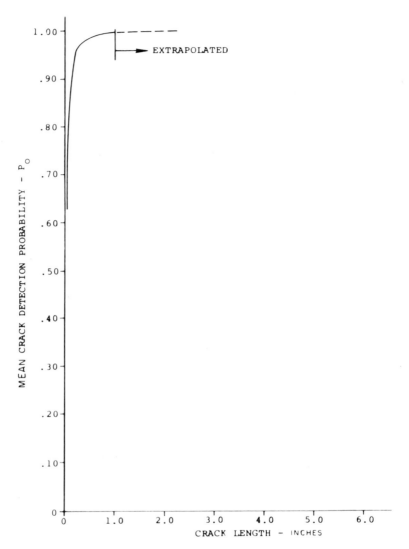

FIG. 8—*Mean crack detection probability for ultrasonic surface wave inspection (Ref 8, Fig. 10-3)*.

The occurrence probability of a maximum stress peak event $S' \geq S(X + \Delta X)$ is $F_E[S(X + \Delta X)]$. Since these events are independent, the probability of occurrence of the combination is the product of the occurrence probabilities for each separate event. Since this combined event also represents an instance of structural failure, the element of structural failure probability is

$$\Delta P_F(X, T) = F_E[S(X + \Delta X)] \cdot \{F_{EN}(a_i[(X + \Delta X), T]) - F_{EN}[a_i(X, T)]\} \quad (14)$$

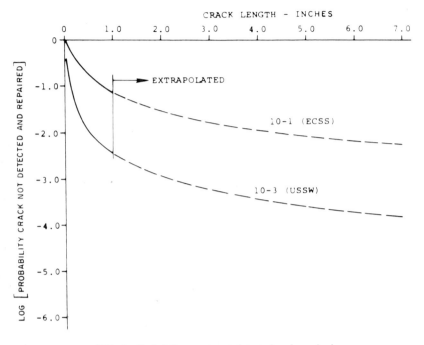

FIG. 9—*Probability crack not detected and repaired.*

And the member failure probability per stress peak event at T mission hours is

$$P_F(T) = \sum_{X_{max}}^{0} \Delta P_F(X, T) \qquad (15)$$

The failure probability per flight of duration ΔT at T mission hours is

$$P_F(T, \Delta T) = \Delta T \cdot (N_0/h) \cdot P_F(T) \qquad [P_F(T) \leq 10^{-7}] \qquad (16)$$

or

$$P_F(T, \Delta T) = 1 - [1 - P_F(T)]^{T \cdot N_0/h} \qquad [P_F(T) > 10^{-7}] \qquad (17)$$

Computation Routine—The computation is carried out with an input data matrix prepared as follows (see Table 2):

1. The largest crack size required as discrete input, X_{max}, is determined as the critical crack length for the stress $_{min}S_{max}$. This value is the head entry in a column of (independently variable) crack lengths at failure, ranging down to zero. Intermediate values are generated by decrementing the previous value by

TABLE 2—Organization of failure probability computation.

Crack Length at Failure, in.	Failing Stress, $S(x)$, ksi	Stress Exceedance Probability, F_E, $S(x)$	T_1, h	T_2, h		Initial Flaw Size, $a_i(X,T)$, in. T_j, h
$X_1 = X_{max}$	$\min S_{max}$	1.00				$a_i(X_1, T_j)$
X_2						$a_i(X_2, T_j)$
X_3						$a_i(X_3, T_j)$
X_i	$S(x_i)$	$F_E[S(x_i)]$				$a_i(X_i, T_j)$
0	F_{tu}	$F_E(F_{tu})$	0	0	0	0

NOTE—$N_0/h =$

$P_F(T) =$ $P_F(T_j)$

amounts that decrease with decreasing crack lengths. (Refer to Table 1 for guidance.)

2. A second column contains as entries the critical stress level for each crack length entry, ranging from $_{min}S_{max}$ at the top to the ultimate strength, F_{tu}, at $X = 0$.

3. A third column contains as entries the exceedance probability for each stress level entered in the second column.

4. The remainder of the table is a flaw/crack size matrix $a_i(X, T)$ with the failing crack lengths, X, as row arguments and instants of service mission time, T, as column arguments. The matrix elements (entries) are generated from Table 1 by locating the growth time to the row argument crack size, then back-calculating the crack growth an amount of time equal to the column argument time, and reading or interpolating the corresponding crack size. Then this is the flaw/crack size that in the column argument time would grow to the row argument crack size. The bottom line entries are all 0.00.

The failure probability computations are carried out separately for each column using Eqs 8, 13, 14, 15, 16, or 17 in turn with the appropriate values for A_k^0, B_k^0 (in Eq 8). Note that the first line in each column is evaluated with $a_i[(X + \Delta X), T] = a_i(X_{max}, T)$ and $F_{EN}(a_i, T) = 0$ (Eq 13). For the last line, Eq 8 is set $= 1.0$, since the probability of exceeding the flaw size 0.00 is unity.

Computation of Inspection/Repair Effects

Analysis—Effects of inspection/repair activities are reflected as changes in the initial equivalent flaw probability distribution function. The experience so far has been that these changes are completely compatible with the form of analytical representation given by Eq 8, and that they can be accommodated by changes in the coefficients A_k^0, B_k^0. The new coefficients will be denoted A_k^j, B_k^j ($j = 1, 2, 3 \ldots =$ number of successive inspection/repair events).

The inspection/repair process operates on the crack sizes, X, listed in the first column in Table 2. When a crack is detected and repaired at time T, it is logically equivalent to removing the flaw size that existed T hours earlier and that grew into the crack, X. Removal of flaws changes the probability distribution of flaw sizes from what it was before the inspection/repair.

The occurrence probability of a largest crack in the size range X, $X + \Delta X$ at time T is in accordance with Eq 13.

Invoking the frequency interpretation of probability leads to an expected number of occurrences of cracks in that size range

$$N_E(X, \Delta X) = N \cdot \Delta P(X, \Delta X) \tag{18}$$

Of this number, a fraction $P_{\text{fix}}(X)$ will be restored to zero time status by repair action, leaving a number

$$N_s(X, \Delta X) = N \cdot \Delta P(X, \Delta X) \cdot (1 - P_{\text{fix}}(X))$$
$$N_s(X, \Delta X) = N \cdot \Delta P(X, \Delta X) \cdot P_s(X) \tag{19}$$

surviving for continued growth. The probability of this event is

$$P_{s_I} = N_s(X, \Delta X)/N = \Delta P(X, \Delta X) \cdot P_s(X) \tag{20}$$

The exceedance probability distribution values are obtained by summation as

$$F_{EN}^{\,j}(a_i, T) = \sum_{a_i(X_{\max}, T)}^{a_i(X, T)} \Delta P(X, \Delta X) \cdot P_s(X) \tag{21}$$

Applying the transformations in Eqs 4 and 5 at each step in the summation, the new coefficients $A_k^{\,j}$, $B_k^{\,j}$ can be evaluated by linear regression of U on V.

Computation Routine—The computations are carried out using an input data matrix somewhat similar to the one used for failure probability computations (Table 3).

1. The row arguments are the final crack sizes used as row arguments in the failure probability computations, omitting the entry 0.00.
2. The second column contains the survival probabilities $P_s(X)$ evaluated by Eqs 9 and 12 for the row argument values of X.
3. The column arguments are the time, T, in mission hours when inspection/repair actions are performed, each with three sub-columns

   (a) The first is the flaw/crack size, $a_i(X, T)$, with elements copied from the failure computation matrix.
   (b) The second is headed by the U_i transformation (Eq 4) with entries to be computed from the progressive summations of Eq 21.

TABLE 3—*Organization of inspection effect computation.*

Crack Length at Failure, X in.	Crack Survival Probability, $P_s(X)$	Mission Time at Inspection						
		T_1				T_j		
		$a_i(X_1 T_1)$	U_{i_1}	V_{i_1}		$a_i(x_1 T_j)$	$U_{i_1 j}$	$V_{i,j}$
$X_1 = X_{max}$								
X_2								
X_3								
X_i	$P_s(X_i)$					$a_i(X_j, T_j)$	$U_{i,j}$	$V_{i,j}$
X_{min} (>0)								

(c) The third is headed by the V_i transformation, Eq 5, with entries being values of that transformation performed on the $a_i(X, T)$ entries.

The computation for the *j*th inspection/repair event is performed by evaluating Eqs 8 (with A_k^{j-1}, B_k^{j-1}), 13, 20, 21, and 4 in turn with entries in the U_i column. The initial sequence is evaluated with $a_i[(X + \Delta X), T] = a_i(X_{max}, T)$ and $F_{EN}(a_i, T) = 0$ (Eq 13). A plot of the U_i versus V_i values helps identify the sets of data pairs to be used in the following regression analysis necessary to determine the segment coefficients, A_k^j, B_k^j, along with the intercept flaw sizes, $a_{i_{k,k+1}}$.

Results and Discussions

Sample calculations were performed to demonstrate the following: general variation of the failure probability with mission time exposure, sensitivity of the failure probability to variations in mission spectrum severity, and effects of periodic inspection/repair actions.

General Trends

Figure 10 shows the general trend of risk variation with operational exposure. The risk (failure probability) increases very slowly at first, but eventually a point is reached where the failure probability starts to increase quite rapidly with increasing service exposure. The onset of the rapid deterioration of the failure probability, while it generally occurs at a comfortably low probability level, could be interpreted as an indication that the useful service life of that member is approaching its end.

Service Exposure Severity

Three failure probability curves are shown in Fig. 10. The middle one is the base-line, generated with the base-line stress exceedance curve shown fully

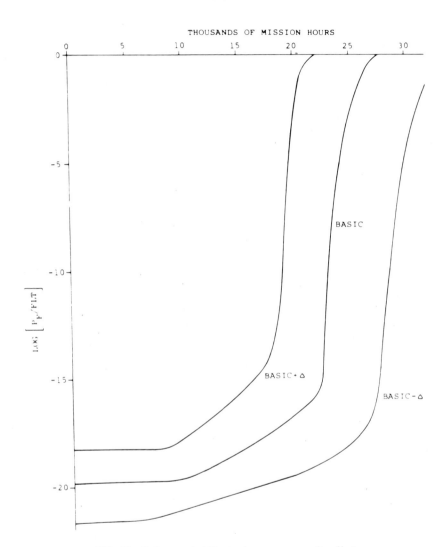

FIG. 10—*Failure probability-load spectrum severity effects.*

drawn in Fig. 1 along with the crack growth curve shown in Fig. 3. The two bracketing curves in Fig. 10 were generated using the dashed stress exceedance curves in Fig. 1 along with appropriately adjusted crack growth data.
The stress exceedance perturbations were obtained by the expression

$$S' = S \pm (0.10) \cdot (S - {}_{min}S_{max}) \qquad (22)$$

that corresponds to an average value for the maximum stress perturbation ratio

$$\overline{S'/S} = 1.0 \pm (0.10) \{1 - [_{min}S_{max}/(F_{tu} - {_{min}}S_{max})] \cdot \ell n(F_{tu}/_{min}S_{max})\} \quad (23)$$

The inverse average stress ratio raised to the third power was used to estimate the effect on the crack growth times of the maximum stress perturbations in Eq 22.

Since the stress spectrum severity is controlled both by the loads environment and by the design decision allocating the 1.0 g gross area stress level, the results are useful in providing exchange ratios between increments of service life extension and increments in design gross area stress levels.

Effects of Periodic Inspections and Repairs

The data shown in the preceeding Figs. 7, 8, and 9 were used to generate data for the periodic inspection effects on failure probability (base-line case). Table 4 showing the proportion of cracks that eludes detection, affords a direct comparison of the effectiveness of the two inspection methods.

Results of the failure probability computations are shown in Figs. 11 through 13. These failure probabilities apply to inspections performed on one airplane. However, the analysis could be extended to apply to a fleet of air-

TABLE 4—*Crack survival comparison.*

	Surviving Fraction	
	Inspection Method	
Crack Size, in.	Eddy Current Surface Scan	Ultrasonic Surface Wave
0.050	0.99	0.37
0.100	0.84	0.14
0.150	0.65	0.075
0.200	0.51	0.048
0.250	0.41	0.034
0.500	0.18	0.011
0.750	0.11	0.006
1.000	0.075	0.0036
1.500	0.044	0.002
2.000	0.030	0.001
2.500	0.020	...
3.000	0.017	...[a]
4.000	0.012	...
5.000	0.008	...
6.000	0.007	...
7.000	0.005	...

[a] Order of 10^{-4}.

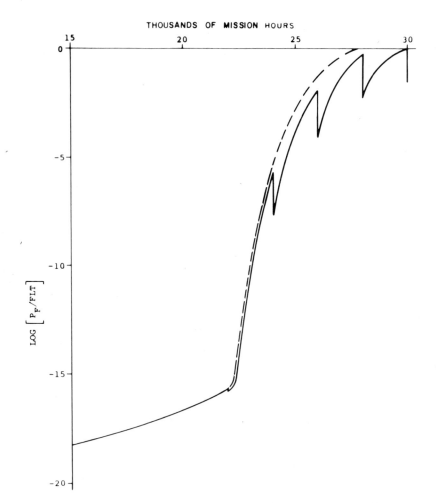

FIG. 11—*Failure probability with periodic inspections. Inspection method, eddy current surface scan [8]; period = 2000 h.*

planes. The first inspection is assumed at 22 000 mission hours. The effect of this inspection on the failure probability is small, because at that time the cracks have not grown much beyond the detection thresholds.

The effects of improved detection effectiveness are clearly evident by comparison of Figs. 11 and 12. An interesting feature is displayed in Fig. 12, indicating that with repeated inspections the preinspection failure probability eventually may peak out, and then start to decline. A similar trend appears present in Fig. 13, although in this case the computations were not carried out across the "hump."

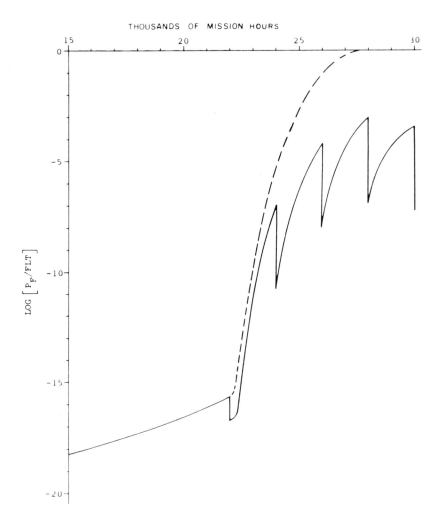

FIG. 12—*Failure probability with periodic inspections. Inspection method, ultrasonic surface wave [8]; period = 2000 h.*

Figure 13 was computed with the same data used to generate Fig. 11, except that the inspection period was reduced from 2000 to 1000 h. Comparison of Figs 11 and 13 illustrates how inspection efficiency can be traded against inspection period length, with attendant changes in inspection burden.

Concluding Remarks

In application of this analysis to a complete airplane, it is necessary to start with a survey to identify the set of structural members that are susceptible to

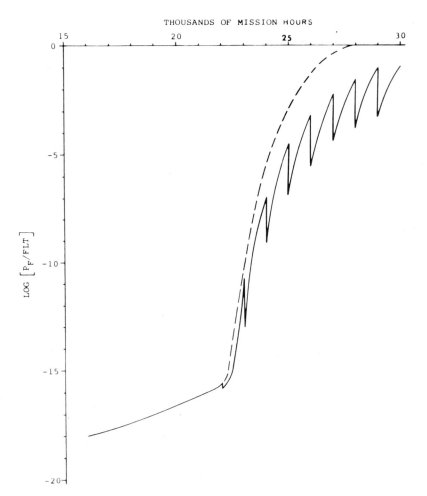

FIG. 13—*Failure probability with periodic inspections. Inspection method, eddy current surface scan [8]; period = 1000 h.*

critical failures from an airworthiness standpoint, with subsequent analysis of each member in the set.

The subject of including structural failure probability considerations as part of structural design criteria is controversial for these reasons:

The computations are generally based on force averages for the principal input parameters, such as stress exceedance and crack growth.

Recent investigations have shown in one case (Ref 6) that approximately 1500 h of operations are needed to achieve statistical stability in these parameters; an individual flight may exceed the statistical average by a factor of 15.

The general public has a tendency to associate the concept of "possibility" with probability, regardless of the numerical value attached to the latter in specific cases.

Since nobody wants to create the impression of the possibility of a structural failure in the products delivered, there is an understandable reluctance to employ terminology that might do just that.

There is on the other hand a growing realization that deterministic criteria (many of which have been derived from statistical considerations) may not be sufficient to estimate the performance of structures over extended periods of time when exposed to random service loading conditions. This leads to the need for complementing the deterministic criteria with some probabilistic considerations, such as the type just outlined.

In application of these considerations, it is necessary to establish tolerable probability levels, however, and therein lies the nub. This paper does not advocate any specific value for a tolerable failure probability. It does include Fig. 14 (from Ref 1) that reflects current thoughts on the matter. This figure relates the apparently acceptable occurrence probabilities of failure events with consequences of these failures.

To circumvent the objection that the analysis is valid for average operations only, the following steps can be taken:

(a) Establish the expected size of the subset of flights performed at consistent above-average operational severity levels.
(b) Determine with pre-selected confidence (90, 95, 99 percent) the severity level appropriate to this subset.
(c) Compute for selected points along the flight time-failure probability curve the failure probabilities at the conclusion of the subset.

When the values are plotted on a graph such as Fig. 10, they will generate an envelope curve that represents the excursions in failure probability with associated confidence.

Finally, it should be mentioned that in the implementation of this type of analysis, it is highly desirable to have it complemented by an individual airplane usage tracking program. The reasons are:

(a) A tracking program provides evidence regarding the conformity of the operating environment with the criteria established for structural design.
(b) It provides a basis for required maintenance actions.
(c) It provides a basis for force management decisions such as route or mission reassignment of individual airplanes.

Acknowledgments

The author wishes to thank H. Bard Allison, Director of Engineering, Lockheed Georgia Company, for permission to publish this paper. Thanks are

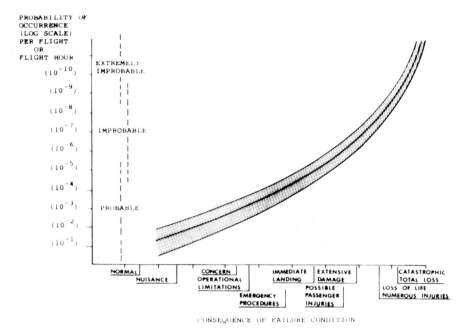

FIG. 14—*Relationship between the consequence of failure and the probability of occurrence.*

also due J. L. Russ, Jr., and F. M. Conley, group engineers, Lockheed Georgia Company, for helpful reviews and discussions in the course of preparing the paper.

References

[1] FAA Advisory Circular AC No. 20 (draft), Federal Aviation Administration, Aug. 1978.
[2] FAA Advisory Circular AC 25.571-1, Federal Aviation Administration, 28 Sept. 1978.
[3] Freudenthal, A. M., "Reliability Assessment of Aircraft Structures Based on Probabilistic Interpretation of the Scatter Factor," AFML-TR-74-198, Air Force Materials Laboratory, April 1975.
[4] Shimozuka, M., "Development of Reliability-Based Aircraft Safety Criteria: An Impact Analysis," AFFDL-TR-76-36, Air Force Flight Dynamics Laboratory, Vol. I, 15 April 1976.
[5] Shinozuka, M., "Reliability-Based Scatter Factors, Volume I: Theoretical and Empirical Results," AFFDL-TR-78-17, Air Force Flight Dynamics Laboratory, Vol. I, March 1978.
[6] Circle, R. L. and Conley, F. M., "Quantitative Analysis of the Variables Involved in Crack Propagation Analysis for In-Service Aircraft," AIAA Paper #80-0752, American Institute of Aeronautics and Astronautics, May 1980.
[7] Gumbel, E. J. "Statistical Theory of Extreme Values and Some Practical Applications," National Bureau of Standards Applied Mathematics Series 33, 12 Feb. 1954.
[8] Lewis, W. H., Sproat, W. H., Dodd, B. D., and Hamilton, J. M., "Reliability of Non-Destructive Inspections-Final Report," San Antonio Air Logistics Center, Report No. 76-6-38-1, Dec. 1978.
[9] Chilver, A. H. "The Frequency of Failures Within a Group of Similar Structures," *The Aeronautical Quarterly*, Nov. 1954.

Pedro Albrecht[1]

S-N Fatigue Reliability Analysis of Highway Bridges

REFERENCE: Albrecht, P., "*S-N* **Fatigue Reliability Analysis of Highway Bridges**," *Probabilistic Fracture Mechanics and Fatigue Methods: Applications for Structural Design and Maintenance, ASTM STP 798*, J. M. Bloom and J. C. Ekvall, Eds., American Society for Testing and Materials, 1983, pp. 184–204.

ABSTRACT: This paper presents a method of calculating the expected fatigue failure probability of a structural detail, given the distribution of resistance and load. The resistance data, in terms of cycles to failure, come from previous laboratory tests. The load data come either from stress range histograms recorded on bridges or from loadmeter surveys. The proposed method replaces each histogram by an equivalent stress range and converts the latter into a distribution in terms of number of cycles. The problem is thus cast into the standard format for reliability analysis and allows one to calculate failure probabilities. Application of the method to designs in accordance with the AASHTO Specifications showed that fatigue failure probabilities for redundant load path (RLP) structures are inconsistent and vary greatly from $P_F = 9.2 \times 10^{-2}$ for Category B to $P_F = 9.2 \times 10^{-10}$ for Category E'. For nonredundant load path (NRLP) structures, they vary from $P_F = 5.1 \times 10^{-2}$ for Category A to $P_F = 2.1 \times 10^{-22}$ for Category E. It is proposed that the specifications be revised to include: (1) allowable stress ranges for RLP and NRLP structures with uniform failure probabilities; (2) explicit formulation of the specifications in terms of the actual number of single "fatigue trucks," each causing an equivalent stress range; and (3) continuous definition of allowable stress range versus truck traffic volume. An example illustrates the design of a bridge, not covered by the AASHTO specifications, to a specified failure probability.

KEY WORDS: fatigue (materials), reliability, fracture mechanics, bridges, steel, probabilistic fracture mechanics

Design methods based on statistical reliability concepts have recently been developed for many areas of static design of members and connections. Code-writing bodies are now incorporating them into their specifications to ensure consistent reliability throughout the structure. Still lacking is a reliability method for fatigue that can then be used as a basis for a load and resistance factor approach to fatigue design. The present study addresses this need.

---

[1] Professor of Civil Engineering, Department of Civil Engineering, University of Maryland, College Park, Md. 20742.

The paper briefly reviews the equivalent stress range and the reliability concepts needed herein. Thereafter, the load and resistance curves are constructed and transformed in a manner suitable for writing the governing equations. The failure probability of designs to the AASHTO fatigue specifications are assessed, and an illustrative design example is presented for a special bridge not covered by the AASHTO specifications [1].[2] The work of Ref 2 is extended to cover designs for "over 2 000 000 cycles" and nonredundant load path structures.

Equivalent Stress Range

Recent studies have employed, with good success, the concept of an equivalent stress range to correlate data from variable amplitude cyclic load tests with data from constant amplitude tests. The concept states that, for equal number of cycles, the equivalent (constant amplitude) stress range will cause the same fatigue damage as the sequence of variable amplitude stress ranges it replaces. For convenience of applying the concept later in the text, the equivalent stress range is expressed in the following form [2]

$$f_{re} = [\Sigma \gamma_i (\phi_i \alpha f_{rd})^m]^{1/m} \tag{1}$$

where

f_{rd} = computed stress range corresponding to the design load;
γ_i = frequency of occurrence of i-th stress range;
ϕ_i = ratio of an individual load to the design load, or ratio of corresponding stress ranges;
α = ratio of measured-to-computed stress range for the design load; and
m = slope of S-N curve.

Figure 1 illustrates for a typical stress range histogram the meaning of the parameters in Eq 1.

One can show mathematically that the equivalent stress range concept and Miner's cumulative damage criteria, when used in conjunction with S-N plots, are special cases of the fracture mechanics approach that is based on a Paris-type equation of crack growth. All three give identical variable amplitude life predictions provided that: (1) the crack initiation phase is negligible; (2) there are no interaction effects in high-low stress range sequences; (3) all stresses are above the constant amplitude fatigue limit; and (4) the inverse slope of the S-N curve and the slope of the crack growth rate curve are equal, that is, $n \cong 3$ for ferritic steels. Experimental data indicate that the equivalent stress range concept works well for bridge load histories even when the four assumptions are only approximately satisfied.

---

[2]The italic numbers in brackets refer to the list of references appended to this paper.

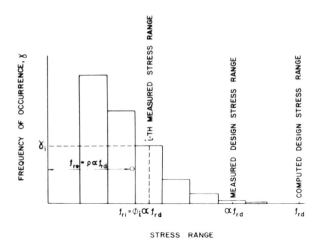

FIG. 1—*Illustration of the terms in Eqs 1 and 3. (Note: if detail is designed to the allowable stress range, then* $f_{rd} = F_{sr}$.)

Since f_{rd} and α are constant for a given stress range histogram, they can be taken out of the summation so that Eq 1 becomes

$$f_{re} = [\Sigma \gamma_i \phi_i^m]^{1/m} \alpha f_{rd} \qquad (2)$$

or

$$f_{re} = \rho \alpha f_{rd} \qquad (3)$$

where ρ is defined in this paper as

$$\rho = [\Sigma \gamma_i \phi_i^m]^{1/m} \qquad (4)$$

The equivalent stress range concept is needed in calculations of fatigue failure probability of structures subjected to variable amplitude stress cycling.

Basic Reliability Concepts

The reliability concepts employed in this paper are well documented in the literature. See, for example, Ref 3.

Structural reliability can be defined as the probability that a structural component will not fail within its design life. In other words, it is the probability that a member's resistance to load is higher than the applied load. In deterministic design, one assumes a high value of load and a low value of resistance and specifies that the distance between the two shall not be less than a preselected safety factor. In probabilistic design, one recognizes that neither the resistance, R, nor the load, Q, are single valued; both have a mean and a distribution. The objective is then to compute the probability of failure, that is, the likelihood of the undesirable cases where a high value of load will exceed a low value of resistance. Conversely, for purposes of writing fatigue design specifications, one wishes to determine the distance between the mean resistance and the mean load, $\tau = \bar{R} - \bar{Q}$, so that the failure probability, P_F, does not exceed a specified value. This is done with the equation for the safety index, β

$$\beta = \frac{\tau}{s_\tau} = \frac{\bar{R} - \bar{Q}}{\sqrt{s_R^2 + s_Q^2}} \tag{5}$$

in which β is the number of standard deviations, s_τ, for the difference between mean load and mean resistance. If load and resistance are normally distributed, so is their difference; and the value of β corresponding to a specified failure probability can be read from tables for the standard normal variable.

Note that increasing the safety index, β, will decrease the failure probability. This can be achieved either by moving the mean load, $\bar{Q}$, farther away from the mean resistance, $\bar{R}$, or by reducing the standard deviation. The second option is usually not available in most designs.

When applying Eq 5 to fatigue design, the resistance is given by the number of cycles to failure and the load is given by the applied stress range history. This leads to two difficulties. One is the need to find a mean and standard deviation of many stress range histograms, each of which describes in itself a distribution of stress ranges. Secondly, the resistance data, which consist of the number of cycles to failure, are distributed along a horizontal line in a S-N plot, but the stress range data are distributed along a vertical line. The following solution to the two difficulties is proposed: (1) replace each histogram by one equivalent constant amplitude stress ranges; (2) calculate the mean and standard deviation of all equivalent stress range; and (3) convert the resulting distribution of equivalent stress ranges into one given in terms of number of cycles. The problem is then reduced to the form to which Eq 5 applies.

Resistance Curve

The AASHTO fatigue specifications [1] are based on constant amplitude fatigue test data for steel beams [4,5]. The statistical analysis of the S-N data has shown that the mean regression line with the best fit was of the log-log linear form

$$\log N = b - m \log f_r \tag{6}$$

with the intercept, b [at $f_r = 6.89$ MPa (1 ksi)], and the slope, m, as the regression coefficients. The log-log plot of Eq 6 gives a straight S-N line, labeled "resistance" in Fig. 2. The data points were found to be log-normally distributed about the mean regression line with about equal standard deviation at all stress range levels. This held true for all details. Thus, one may assume that for any point, f, on the mean regression line, the fatigue life of replicate specimens tested at the same stress range would be log-normally distributed about that point with mean $\bar{R} = \log N$ and standard deviation.

$$s_R = s_{\log N} \tag{7}$$

The mean and the standard deviation of the number of cycles to failure define the resistance.

The regression coefficients, b and m, and the standard deviation, s_R, for the six categories A through E' are summarized in Table 1. They provide the

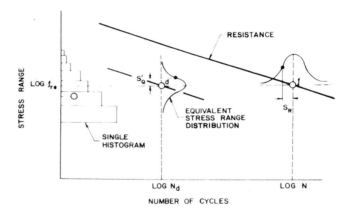

FIG. 2—*Construction of equivalent stress range distribution.*

TABLE 1—*Resistance curve parameters from data in Refs 4 through 7.*

Category	Type of Detail Tested	Number of Data Points	Regression Coefficients		Standard Deviation, s_R
			Intercept, b^a	Slope, m	
A	rolled beam	28	11.121	3.178	0.221
B	welded beam	55	10.870	3.372	0.147
C*	stiffeners	135	10.085	3.097	0.158
C	2-in.^{b} attachments	14	10.0384	3.25	0.0628
D	4-in. attachments	44	9.603	3.071	0.108
E	cover plate end	193	9.2916	3.095	0.1006
E'	cover plate end, $t > 0.8$ in.	18	9.1664	3.2	0.1943

^{a}For values of f_r substituted in units of ksi (1 ksi = 6.89 MPa).
^{b}1 in. = 25.4 mm.

resistance curve parameters needed in Eq 5. The data listed under Category C and E' require some explanation. In the AASHTO specifications, Category C covers both transverse stiffeners and 50-mm (2-in.) attachments. The mean regression line of the former falls higher than that of the latter, making it appear that the 50-mm (2-in.) attachment data would govern. In reality, one must also consider the standard deviation, which is about 2½ times larger for stiffeners than for 50-mm (2-in.) attachments. This creates a peculiar situation. If one moves to the left of each mean by up to 2.56 standard deviations, the 50-mm (2-in.) attachment governs. Beyond that, however, the transverse stiffener data become critical. Accordingly, this study employs that data for Category C that governs at the value of β being considered. In contrast, the AASHTO fatigue specifications are solely based on the 50-mm (2-in.) attachment data, although safety indices for Category C reach a value of 7.12 for nonredundant load-path structures [6].

Table 1 also has an entry for Category E', although no regression analysis was reported in Ref 7, presumably because of a lack of data at various stress range levels in the finite life region of the S-N plot. For the purpose of this study, the finite life region for Category E' details was defined by the 18 data points at 55 MPa (8 ksi) stress range for which the mean life was 1 890 000 cycles, with standard deviation of log of life, $s_R = 0.1943$. Assuming a slope, $m = 3.2$, equal to the mean of the six other slopes in Table 1, the intercept is then given from Eq 6 as

$$b = \log(1.89 \times 10^6) + 3.2 \log(8.0) = 9.1664$$

Load Curve

The load curve data can come either from field measurements of stress range histories or from loadometer surveys. In the former case, the strain ranges caused by the applications of a live load are obtained from a strain gage mounted at a suitable point on the bridge. In the latter case, the trucks are weighed. The results are usually reported as a histogram of stress range or truck weight versus frequency of occurrence.

The proposed construction of the load curve is illustrated below for a class of comparable structures, all of which are subjected to the same design load. For example, the fatigue design of short-span highway bridges is governed by the number of single truck crossings. Each application of the design load induces one stress range cycle. It is assumed that one has available stress range histograms recorded on several bridges.

The construction of the load curve begins with a single stress range histogram, such as the one plotted along the ordinate in Fig. 2. The bar width in a histogram is usually constant; it varies in Fig. 2 because the S-N plot scales are logarithmic. The equivalent stress range, f_{re}, of the single histogram is then calculated. It replaces the histogram in subsequent calculations and provides one point for the desired load curve. Plotting the distribution of all equivalent stress ranges on a vertical line through the design point, d, gives the load curve. To define its distribution, one needs the mean and the standard deviation. Assuming that the ratio of measured-to-computed stress range, α, is constant, implies that αf_{rd} is also constant. Therefore, the computation of the log mean of all equivalent stress ranges is reduced to evaluating the log-mean of ρ for all histograms (see Eq 3).

$$\overline{\log \rho} = \frac{1}{h} \Sigma \log \frac{f_{re}}{\alpha f_{rd}} \qquad (8)$$

where h is the number of histograms. The standard deviation of the load is then given by the standard deviation of the log ρ values.

$$s'_Q = s_{\log \rho} \qquad (9)$$

The prime added to s'_Q and to any other symbol indicates a quantity measured along a vertical line.

The load is defined by the line through the design point, d, drawn parallel to the resistance, and the standard deviation of the equivalent stress ranges. The load curve could be derived in analogous fashion from the results of loadometer surveys if one assumes that loads are proportional to stresses.

All bridges on the state and federal highway systems are designed for the same load history that was derived from a nationwide loadometer survey. Estimation of the load curve is less certain when the structure is one of a kind and few data are available. In that case, one must construct the load histogram expected over the design life, compute the equivalent stress range for that histogram, and estimate the standard deviation.

Transformation of Load and Resistance

Equation 5 applies only if the load and resistance curves are plotted side by side with the same base line. The load is distributed along a vertical line through the design point, d, shown in Fig. 2, whereas the resistance is distributed along a horizontal line through the failure point, f. One of the two curves must, therefore, be transformed.

Reference 2 presented the transformation of the load curve, whereas this paper explains instead the transformation of the resistance curve in terms of self-evident geometrical relationships. Figure 3 shows the solid resistance line and two dashed lines shifted above and below the mean resistance by a specified number of standard deviations, so that the survival probability (assuming a single-valued load) would be, say, 5 percent and 95 percent, respectively. The distribution of the resistance is drawn along a horizontal line through Point f and also along a vertical line through Point g. In both cases,

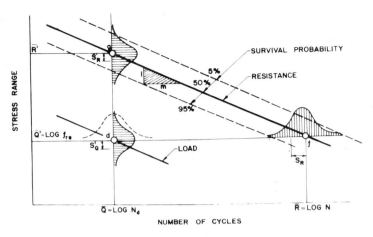

FIG. 3—*Transformation of load and resistance.*

the points with the same survival probability must lie on the same line parallel to the mean resistance. Since the slope is $1:m$, it follows for reasons of geometry that

$$s'_R = \frac{s_R}{m} \qquad (10)$$

where s_R and s'_R are the standard deviations of the resistance when its distribution is drawn about a horizontal and a vertical line, respectively. For the same geometrical reasons, the distance between the mean resistance and the mean load, measured along the vertical line g-d, is given by

$$\bar{R}' - \bar{Q}' = \frac{1}{m}(\bar{R} - \bar{Q}) \qquad (11)$$

Rewriting Eq 5 for the distance g-d, and substituting Eq 10 and 11 gives

$$\beta = \frac{\bar{R}' - \bar{Q}'}{\sqrt{(s'_R)^2 + (s'_Q)^2}} = \frac{(1/m)(\bar{R} - \bar{Q})}{\sqrt{(s_R/m)^2 + (s'_Q)^2}} \qquad (12)$$

or, after simplifying

$$\beta = \frac{\bar{R} - \bar{Q}}{\sqrt{(s_R)^2 + (ms'_Q)^2}} \qquad (13)$$

Equation 12 applies to distributions of load and resistance along a vertical line through Points d and g, respectively. The required 90-deg rotation of the resistance distribution decreases the standard deviation by a factor, m, in accordance with Eq 10. The load remains unchanged.

Equation 13 applies to distributions along a horizontal line through points d and f. In this case, the load distribution is rotated by 90 deg, thus increasing its standard deviation to $s_Q = m s'_Q$, compatible with the geometrical condition stated in Eq 10. The resistance remains unchanged. Both equations give the same result for β.

Finally, substituting the identities $\bar{R} = \log N$ and $\bar{Q} = \log N_d$ into Eq 13 and using the abbreviation for the standard deviation of the difference between the resistance and the load

$$s_\tau = \sqrt{(s_R)^2 + (m s'_Q)^2} \tag{14}$$

yields the safety index for fatigue design.

$$\beta = \frac{1}{s_\tau} (\log N - \log N_d) \tag{15}$$

The terms s_R and s'_Q are given by Eqs 7 and 9, respectively. The failure probability corresponding to any numerical value of β can be read from tables for the normal distribution.

Equation 15 addresses the problem of computing the failure probability for a given design. The solution to the reverse problem, that of computing the design life, N_d, for a desired failure probability, follows from the same equation.

$$\log N_d = \log N - \beta s_\tau \tag{16}$$

Upon substitution of N from Eq 6, with $f_r = F_{re}$, one gets

$$\log N_d = (b - \beta s_\tau) - m \log F_{re} \tag{17}$$

or, taking the antilog

$$N_d = \frac{10^{(b - \beta s_\tau)}}{F_{re}^m} \tag{18}$$

The mean safety factor on life is, from Eq 16

$$(FS)_N = \frac{N}{N_d} = 10^{\beta s_\tau} \tag{19}$$

and that on stress range

$$(FS)_{f_r} = (FS)_N^{1/m} = 10^{\beta s_\tau / m} \tag{20}$$

For a fixed value of β, the safety factors vary with type of detail because the standard deviation of the resistance, s_R, varies.

In summary, Eqs 15 and 18 yield the safety index and the allowable number of cycles, respectively, for a given design or a desired failure probability.

Assumptions

The assumptions made in this paper are summarized for convenience hereafter. Many are based on data from previous studies and are referenced accordingly; others rely on engineering judgment when few data exist.

Assumptions Pertaining to the Resistance Curve

1. The log-log linear S-N curve for constant amplitude fatigue test data is extended below the constant amplitude fatigue limit, F_L, downwards to a point where the equivalent stress range meets the variable amplitude fatigue limit at [8]

$$F_{Le} = \rho F_L \qquad (21)$$

2. Calculations pertaining to a specific type of detail employ the exponent, m, corresponding to the slope of the S-N curve for that detail. Calculations of equivalent stress range are insensitive to small changes in m. Therefore, a rounded value, $m = 3$, is used, lending to Eq 1 the meaning of a root-mean-cube (RMC) stress range [9].

3. Load interaction effects in high-low stress range sequences are neglected because, in most civil engineering structures, the random nature of loading does not provide enough low load cycles following a high load excursion to retard crack growth [10].

4. The fabrication quality of the test specimens, from which the resistance data of Table 1 originate [4,5,7], is representative of all structures in service.

5. Except for thick cover plates, now covered by the newly adopted Category E' [11], any effect of specimen size and plate thickness on the fatigue life is neglected.

6. Loss of life due to corrosion fatigue (all steels) and weathering (A588 steel) is neglected, although it can be high for high fatigue strength details [12].

7. The resistance data is log-normally distributed [4,5].

Assumptions Pertaining to the Load Curve

8. The maximum stress range in a recorded histogram is caused by the design load.

9. The measured-to-computed stress range ratio, α, has a single value with no distribution.

10. Available loadometer surveys and stress range histograms describe typical load variability for highway bridges.

11. The load data is log-normally distributed [13].

It should be emphasized that the safety index and failure probability, computed from Eq 15, apply to one detail. Since all structures have more than one detail, the probability that the first detail will fail is about equal to the sum of the failure probabilities of all details. Finally, failure of the first detail does not necessarily induce collapse. This depends on the redundancy of the load path.

The user must evaluate the assumptions just listed and the remarks on failure probabilities when he applies the proposed method to a specific problem.

Application to AASHTO Specifications

This section illustrates the first type of application, namely, to compute with Eq 15 the failure probability of a detail that was designed to a specified design life.

Background

The current fatigue specifications for highway bridges, railway bridges, buildings, and weldments state in identical tables the allowable stress range, F_{sr}, as a function of type of detail and number of loading cycles, N_d. (See Table 1.7.2A1, Ref 1.) The listed pairs of stress range versus number of loading cycles are coordinates of points on the allowable S-N lines for each type of detail. For redundant load path structures, these lines were set at two standard deviations, $2s_R$, to the left of the resistance [14]. They are loosely called in the literature the "95 percent confidence limit for 95 percent survival," although a design to those allowable S-N lines will not give a failure probability of $P_F = 5$ percent (95 percent survival) for reasons that will become apparent in the following. It should also be noted that the current specification makes no allowance for load variability. Substituting accordingly $s'_Q = 0$ and $\beta = 2$ into Eq 16, and computing log N from Eq 6 with $f_r = F_{sr}$ gives, indeed, the AASHTO fatigue design lines for redundant load path structures.

$$\log N_d = \log N - 2s_R = (b - 2s_R) - m \log F_{sr} \tag{22}$$

The tabulated values of F_{sr} and N_d are approximate coordinates of points on those lines. Equations 6 and 22 for Category E details are plotted in Fig. 4.

Redundant Load Path Structures

Equation 15 is applied to loading Case I for which the average daily truck traffic equals ADTT = 2500 (or more). It cannot be applied to Case II and

Case III, without assuming a frequency of loading, because AASHTO does not specify the value of ADTT for those cases.

Consider, for example, a Category E cover plate end detail on a RLP structure designed to Point e on the allowable S-N line shown in Fig. 4. Its coordinates are $F_{sr} = 55$ MPa (8 ksi) and N_d (AASHTO) $= 2\,000\,000$ cycles. To locate the actual design point, d, one must find the equivalent stress range, f_{re}, and the actual number of loading cycles, N_d.

The equivalent stress range was extracted from the information reported in Ref *14*. In that report, a linear relationship was assumed between gross vehicle weight and stress range. Accordingly, the coefficients ρ and α that relate the equivalent stress range, f_{re}, to the design stress range, F_{sr}, can be obtained from gross vehicle weight data. The gross vehicle weight distribution from the 1970 Federal Highway Administration (FHWA) nationwide loadometer survey yielded a summation of $\gamma_i \phi_i^3 = 0.35$ [*14*]. This gives, in the manner of Eq 4, $\rho = 0.705$. The ratio of the actual stress range due to the passage of a design vehicle and the design stress range is $\alpha = 0.5$. It is obtained from Eq 6 in Ref *14* for a 50-year design life. Note that F_{sr} is based on a distribution factor for wheel loads to girders on a bridge designed for two or more traffic lanes, $S/5.5$, where S is the girder spacing. The combination of $S/5.5$ with $\alpha = 0.5$ gives $S/11$, a plausible value for the distribution factor for bridges designed for one traffic lane. The equivalent stress range is then

$$f_{re} = 0.5 \times 0.705 \times 8 = 19 \text{ MPa} (2.8 \text{ ksi})$$

For purposes of illustration, a stress range histogram proportional to the gross vehicle weight histogram from the nationwide loadometer survey, was used in Ref *14*, is shown along the ordinate of Fig. 4. (See also Fig. 25 in Ref *14*.) For Loading Case I and ADTT $= 2500$, AASHTO specifies that a Category E detail be designed to the hypothetical point, e, in Fig. 4, for N_d (AASHTO) $= 2\,000\,000$ cycles of the allowable stress range, $F_{sr} = 55$ MPa (8 ksi). The $\alpha F_{sr} = 28$ MPa (4 ksi) stress range corresponds to one 320 kN (72 000 lb) truck on a bridge designed for one traffic lane. The $f_{re} = \alpha \rho F_{sr} = 19$ MPa (2.8 ksi) stress range corresponds to one $\rho \cdot 72\,000 = 226$ kN (50 760 lb) "fatigue truck" on a bridge designed for one traffic lane. (See also the first entry in Table 5 of Ref *15*.) The actual number of cycles for a 50-year design life is $N_d = 2500$ trucks/day $\times 365$ days $\times 50$ years $= 45\,625\,000$ cycles. The actual design point, d, has therefore the coordinates $f_{re} = 19$ MPa (2.8 ksi) and $N_d = 45\,625\,000$ cycles. That the actual design point, d, also lies on the load curve, at $2s_R$ to the left of the resistance, follows from the identities

$$\left(\frac{F_{sr}}{f_{re}}\right)^m = \left(\frac{1}{0.5 \cdot 0.705}\right)^3 = 22.83$$

$$\frac{N_d}{N_d \text{(AASHTO)}} = \frac{45\,625\,000}{2\,000\,000} = 22.81$$

(23)

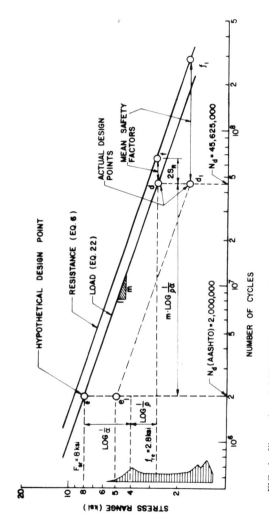

FIG. 4—*Illustration of AASHTO fatigue specification requirements for Category E and ADTT = 2500.*

Equations 23 reflect the geometrical relationship that the slope times the rise must equal the flat of the log-log linear design S-N line.

Evidently, the intentional mismatch between the AASHTO number of cycles of F_{sr} stress range and the actual number of cycles of f_{re} stress range means that the AASHTO specifications apply in reality to one traffic lane loaded by a single "fatigue truck." The fatigue design to the hypothetical point, e, is mathematically identical to a design to the actual point, d. This conclusion is illustrated in Fig. 4 for a Category E detail, but it holds equally for all other categories.

The horizontal distance between the failure point, f, and the actual point, d, is therefore $2s_R$. Substituting this value into Eq 15 gives the safety index for main longitudinal load carrying members in RLP structures designed for ADTT = 2500.

$$\beta = \frac{2s_R}{s_T} \qquad (24)$$

Note again that neglecting load variability implies $s'_Q = 0$ and leads to $\beta = 2$, as in Eq 22. The AASHTO requires that "members shall also be investigated for over two million stress cycles produced by placing a single truck on the bridge distributed to the girders as designated in Article 1.3.1 (B) for one traffic lane loading." (The distribution factor for one traffic lane loading is $S/7$). This requirement lowers the hypothetical design point in Fig. 4 from e to e_1 and the actual design point from d to d_1. It increases the numerator of Eq 24 from the distance $d - f$ to $d_1 - f_1$. The safety index for "over 2 000 000 cycles" is then given by

$$\beta = \frac{1}{s_T} \left[2s_R + m \cdot \log \frac{F_{sr}(2 \times 10^6)}{F_{sr}(\text{over } 2 \times 10^6)} \right] \qquad (25)$$

in which the allowable stress ranges for "2 000 000 cycles" and "over 2 000 000 cycles" are read from AASHTO Table 1.7.2Al for RLP structures. There is really no need to check both loading conditions for every design, since the one that governs can be determined a priori. The condition for "2 000 000 cycles" of loading always governs for Categories A, B, and C* because $F_{sr}(2 \times 10^6)/F_{sr}(\text{over } 2 \times 10^6) < (S/5.5)/(S/7)$. The condition for "over 2 000 000 cycles" always governs for Categories C, D, E, and E' because $F_{sr}(2 \times 10^6)/F_{sr}(\text{over } 2 \times 10^6) > (S/5.5)/(S/7)$. Hence, Eq 24 always applies to Categories A, B, and C*, Eq 25 always applies to Categories C, D, E, and E'. Both are in reality for single truck loading; the former with a distribution factor $\alpha(S/5.5) = S/11$, as shown previously; the latter with $\alpha(S/7) = S/14$. The double check requirement is superfluous and leads to inconsistent failure probabilities.

The numerical evaluation of Eqs 24 and 25 was carried out for all categories. The values of s_R and m, needed to calculate s_τ with Eq 14, are listed in Table 1. Lacking variability information for "fatigue truck" weights, the standard deviation of the load was set equal to the standard deviation of the equivalent stress ranges that were obtained from 104 histograms recorded on 29 bridges in eight states, $s'_Q = 0.0492$ [13]. The allowable stress ranges for main longitudinal load carrying members in RLP structures were taken from AASHTO Table 1.7.2Al [1]. The calculated safety indices and failure probabilities are shown in the left part of Table 2. The results reveal extreme variations in failure probability, ranging from a high of $P_F = 9.2 \times 10^{-2}$ for Category B to a low of $P_F = 9.2 \times 10^{-10}$ for Category E$'$.

Nonredundant Load Path Structures

Failure probabilities of main longitudinal load carrying members in NRLP structures can be calculated in a similar manner. Again, the condition for "2 000 000 cycles" of loading always governs for Categories A, B, and C*, for which the safety index is

$$\beta = \frac{1}{s_\tau} \left[2s_R + m \cdot \log \frac{F_{sr}(2 \times 10^6; \text{RLP})}{F_{sr}(2 \times 10^6; \text{NRLP})} \right] \quad (26)$$

The condition for "over 2 000 000 cycles" of loading always governs for Categories C, D, and E, with safety index of

$$\beta = \frac{1}{s_\tau} \left[2s_R + m \cdot \log \frac{F_{sr}(2 \times 10^6; \text{RLP})}{F_{sr}(\text{over } 2 \times 10^6; \text{NRLP})} \right] \quad (27)$$

The results, shown to the right in Table 2, reveal once more extreme variations in failure probability, from a high of $P_F = 5.1 \times 10^{-2}$ for Category A to a low of $P_F = 2.1 \times 10^{-22}$ for Category E. Note also, that the failure probabilities for Category A, B, and C* details on NRLP structures exceed the failure probabilities for Category C, D, E, and E$'$ details on RLP structures. The original intent of adding a table of F_{sr} values for NRLP structures had been to lower failure probabilities to less than those for RLP structures.

Design Example

This section illustrates with one example the second type of application, namely, the fatigue design of a detail for a desired failure probability. This is done for a special bridge subjected to truck traffic that would not be permitted on public highways. Its design is therefore not covered by the AASHTO specifications. The engineer would have to set his own design criteria in this case.

TABLE 2—*Fatigue failure probabilities for main longitudinal load carrying members designed for ADTT = 2500.*

		Redundant Load Path Structures				Nonredundant Load Path Structures			
	Standard Deviation of $\bar{R} - \bar{Q}$,	2 000 000 Cycles		Over 2 000 000 Cycles		2 000 000 Cycles		Over 2 000 000 Cycles	
Category	s_τ (Eq 14)	β (Eq 24)	P_F	β (Eq 25)	P_F	β (Eq 26)	P_F	β (Eq 27)	P_F
A	0.2707	1.633	5.1×10^{-2}	...	...	1.633	5.1×10^{-2}	...	...
B	0.2217	1.326	9.2×10^{-2}	...	...	2.104	1.8×10^{-2}	...	...
C*	0.2195	1.440	7.5×10^{-2}	...	...	1.930	2.7×10^{-2}	...	...
C	0.1718	...	...	2.887	1.9×10^{-3}	...	...	3.752	8.8×10^{-5}
D	0.1857	...	...	3.725	9.5×10^{-5}	...	...	6.141	4.1×10^{-10}
E	0.1825	...	...	4.564	2.5×10^{-6}	...	...	9.669	2.1×10^{-22}
E'	0.2501	...	...	6.012	9.2×10^{-10}	...	...	...	...

Problem

A one-lane bridge consists of two plate girders and a concrete deck. It is located on a private access road from an ore concentrating plant to the mine. The trucks cross the bridge empty on the way to the mine and full on the way back to the plant. The net and gross vehicle weights are NVW = 40 kips and GVW = 140 kips. Since all other vehicles weigh much less than the empty ore truck, their contribution to fatigue damage is negligible. About 150 daily round trips will keep the plant working at full capacity. At that rate, all ore will be mined in 12 years. Since the bridge has low clearance, crosses a shallow river, and serves no public roads, a low safety index of $\beta = 3$ ($P_F = 1.35 \times 10^{-3}$) is assumed. Compute the allowable stress range for the Category B flange-to-web weld.

Solution

The load and resistance curve data are determined and substituted into Eq 17. Its solution yields the allowable stress range for the specified number of loading cycles. See Fig. 5.

(*a*) Load curve data:

Vehicle weights and frequencies:

$$\phi_{NVW} = \frac{40}{140} = 0.286 \qquad \gamma_{NVW} = 0.5$$

$$\phi_{GVW} = \frac{140}{140} = 1.0 \qquad \gamma_{GVW} = 0.5$$

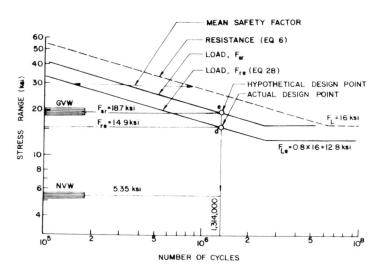

FIG. 5—*Design example.*

Substitute in Eq 4:

$$\rho = [0.5(0.286)^3 + 0.5(1.0)^3]^{1/3} = 0.80$$

For a one-lane bridge assume: $\alpha = 1.0$
A 15-percent coefficient of variation, $C = 0.15$, is estimated for the equivalent truck weight, so that [16]

$$s'_Q = \sqrt{0.4343 \log_{10}(1 + C^2)} = 0.0648$$

(b) Resistance curve data for Category B, from Table 1:

$$b = 10.870 \qquad m = 3.372 \qquad s_R = 0.147$$

(c) Design equation (Eqs 14 and 18):

$$s_\tau = \sqrt{(0.147)^2 + (3.372 \times 0.0648)^2} = 0.2634$$

$$N_d = \frac{10^{(10.870 - 3 \times 0.2634)}}{F_{re}^{3.372}} = \frac{12.0 \times 10^9}{F_{re}^{3.372}} \qquad (28)$$

(d) Number of loading cycles

$$N_d = (2 \times 150 \text{ trips})(365 \text{ days})(12 \text{ years}) = 1\ 314\ 000 \text{ cycles}$$

(e) Allowable stress range, for design based on equivalent truck weight:

$$F_{re} = \left[\frac{12.0 \times 10^9}{1.314 \times 10^6}\right]^{1/3.372} = 103 \text{ MPa (14.9 ksi)}$$

for design based on GVW:

$$F_{sr} = \frac{F_{re}}{\rho \alpha} = \frac{14.9}{0.80 \times 1.0} = 129 \text{ MPa (18.7 ksi)}$$

(f) Check fatigue limit:

$$f_r(\text{GVW}) = 129 \text{ MPa (18.7 ksi)} > F_L = 110 \text{ MPa (16 ksi)}$$

Therefore, fatigue must be checked.
(g) Safety factor on life, from Eq 19:

$$(\text{FS})_{N_d} = 10^{(3 \times 0.2634)} = 6.2$$

The results of the previous calculations are shown in Fig. 5. The histogram is plotted along the ordinate. The upper and lower load lines are for designs to F_{sr} and F_{re}, respectively. Both give analogous results since the two lines are shifted by the ratio $\alpha\rho = F_{re}/F_{sr}$. The safety factor on life is the horizontal distance between the resistance curve and the F_{re} load curve.

Conclusions and Recommendations

A method of calculating fatigue failure probabilities, based on the S-N approach, was presented in a form suitable for examining designs to current fatigue specifications and for special designs to any desired level of risk. It was illustrated herein in detail for highway bridges. The main findings were as follows:

1. The design for the AASHTO number of cycles of maximum stress range, F_{sr}, calculated for multiple HS-20 trucks with a distribution factor for two or more traffic lane loading, $S/5.5$, is mathematically identical to the design for the actual number of single "fatigue trucks" with a distribution factor of $S/11$.
2. The dual requirement to check all Case I designs for "2 000 000 cycles" with $S/5.5$ and for "over 2 000 000 cycles" with $S/7$ is superfluous because the former always governs for Categories A, B, and C*, and the latter always governs for Categories C, D, E, and E'. It is also inconsistent because both are in reality for single "fatigue truck" loading, but the former for a higher stress range with an actual distribution factor, $\alpha(S/5.5) = S/11$, and the latter for the fatigue limit with an actual distribution factor $\alpha(S/7) = S/14$.
3. The failure probabilities lack uniformity. For RLP structures, they vary from a high of $P_F = 9.2 \times 10^{-2}$ for Category B to a low of $P_F = 9.2 \times 10^{-10}$ for Category E'. For NRLP structures, they vary from a high of $P_F = 5.1 \times 10^{-2}$ for Category A a low of $P_F = 2.1 \times 10^{-22}$ for Category E.
4. The failure probabilities for Category A, B, and C* details on NRLP structures are higher than those for Category C, D, E, and E' details on RLP structures. This violates the intent of the requirements for NRLP structures.

In general, one should be careful not to interpret calculated failure probabilities as absolute values. The accuracy of the results always depends on the assumptions inherent in the mathematical model and the available data base. The greatest benefits of such calculations are the ability to: (1) examine the results in terms of the relative failure probability of one design versus another, and (2) set new design criteria based on uniform failure probabilities and experience with similar structures.

The need for specifications that are based on uniform failure probabilities suggests that AASHTO choose two safety indices, one for RLP and one for NRLP structures. Thereafter, allowable stress ranges can be derived with the method outlined herein. In addition AASHTO should consider making the following revisions that have also been suggested by other investigators:

1. Define the allowable stress range as a continuous function of truck traffic volume, instead of the step function approach by Loading Case in Table 1.7.2B.
2. Explicitly formulate the fatigue specifications in terms of the actual number of single "fatigue trucks," each causing an equivalent stress range, instead of a hypothetical number of HS-20 design trucks causing the maximum stress range.
3. Examine what impact the distribution factors would have on the fatigue specifications if they were expressed in terms of number of girders and lanes, instead of girder spacing.

References

[1] *Standard Specifications for Highway Bridges*, 12th ed, American Association of State Highway and Transportation Officials, Washington, D.C., 1977.
[2] Albrecht, P., "Analysis of Fatigue Reliability," Civil Engineering Report, University of Maryland, College Park, Md., Jan. 1981.
[3] Galambos, T. V., "Proposed Criteria for Load and Resistance Factor Design of Steel Building Structures," Research Report No. 45, Civil Engineering Department, Washington University, St. Louis, Mo., May 1976.
[4] Fisher, J. W., Frank, K. H., Hirt, M. A., and McNamee, B. M., "Effects of Weldments on the Fatigue Strength of Steel Beams," NCHRP Report No. 102, Highway Research Board, National Academy of Science, National Research Council, Washington, D.C., 1970.
[5] Fisher, J. W., Albrecht, P. A., Yen, B. T., Klingerman, D. J., and McNamee B. M., "Fatigue Strength of Steel Beams with Welded Stiffeners and Attachments," NCHRP Report No. 147, Highway Research Board, National Academy of Science, National Research Council, Washington, D.C., 1974.
[6] Albrecht, P. and Simon, S., "Fatigue Notch Factors for Structural Details," *Journal of the Structural Division*, American Society of Civil Engineers, July 1981.
[7] Fisher, J. W., Hausammann, H., and Pense, A. W., "Retrofitting Procedures for Fatigue Damaged Full-Scale Welded Bridge Beams," Report No. 417-3 (79), Fritz Engineering Laboratory, Lehigh University, Bethlehem, Pa., Jan. 1979.
[8] Albrecht, P. A. and Friedland, I. M., "Effect of Fatigue Limit on Variable Amplitude Cyclic Behavior of Stiffeners," *Journal of the Structural Division*, American Society of Civil Engineers, Dec. 1979.
[9] Yamada, K. and Albrecht, P., "Fatigue Design of Welded Bridge Details for Service Stresses," Transportation Research Record No. 607, Transportation Research Board-National Research Council, Washington, D.C., April 1977.
[10] Albrecht, P. and Yamada, K., "Simulation of Service Fatigue Loads for Short-Span Highway Bridges," *Service Fatigue Loads Monitoring, Simulations, and Analysis, ASTM STP 671*, American Society for Testing and Materials, Philadelphia, Pa., 1979.
[11] "Interim Specifications Bridges," American Association of State Highway and Transportation Officials, Washington, D.C. 1980.
[12] Albrecht, P., "Fatigue Behavior of 4-Year Weathered A588 Steel Specimens with Stiffeners and Attachments," Report No. FHWA/MD-81/02, Department of Civil Engineering, University of Maryland, College Park, Md., July 1981.

[13] Duerling, K., "Guidelines for Variable Amplitude Fatigue Design Based on Reliability", Masters thesis, University of Maryland, College Park, Md., 1978.
[14] Fisher, J. W., "Bridge Fatigue Guide," American Institute of Steel Construction, New York, N.Y., 1977.
[15] Committee on Loads and Forces on Bridges, "Recommended Design Loads for Bridges," *Journal of the Structural Division*, American Society of Civil Engineers, July 1981.
[16] Wirsching, R. H., "Probability-Based Fatigue Design Criteria for Offshore Structures," University of Arizona, Tucson, Ariz., Feb. 1981.

Summary

Statistical Representation and Methodologies

The order of papers and subheadings in this book reflect the original organization of the symposium program into sessions on Probabilistic Fracture Mechanics and Statistical Aspects of Fatigue. However, there is an overlapping of subject matter within these headings, and the papers could be also viewed as covering two somewhat different categories: (1) the statistical representation of one or more variables affecting structural safety or structural reliability and (2) statistical methodology for considering the interrelationship of variables affecting structural reliability and safety. In the first category are papers by Johnston, Trantina and Johnson, Berens and Hovey, Ostergaard and Hillberry, and Larson and Shawver. The other six papers belong to the second category. All the papers demonstrate the applicability of their approaches and cover several structural applications including pressurized piping for nuclear power plants, aerospace structures, and bridge structures.

In the first category on statistical representation, *Johnston* directs us to the importance of knowing the statistical scatter of both fracture toughness and fatigue crack growth rates. *Ostergaard and Hillberry* discuss how to statistically characterize one variable, the fatigue crack growth properties of a material. *Trantina and Johnson's* study shows that S-N fatigue data can be combined with fatigue crack growth/linear elastic fracture mechanics models to obtain a probabilistic distribution of critical initial defects. *Berens and Hovey* address the important question of the determination of inherent uncertainties in nondestructive evaluation using a probability of crack detection as a function of crack size. Actual data from a recent Air Force study on nondestructive evaluation (NDE) reliability is used to estimate the parameters of the selected NDE model. Finally, *Larson and Shawver* present a probabilistic method of calculating accumulated fatigue damage at an aircraft structural location which accounts for variation in fatigue loads due to Mach number, altitude, gross weight, and material fatigue properties.

In the second category on statistical methodology, *Jouris* emphasizes failure probabilities compared to existing deterministic ASME Code margins of safety. *Harris and Lim* consider the influence of the importance of in-service inspection on structural reliability. Attention is directed to both initial crack distributions and subsequent crack distributions relevant to the quality of in-service inspection. *Walker* presents a relatively simple probabilistic model incorporating fatigue to a given initial crack size, crack growth, probability of crack detection and a residual strength safety criterion. Bronn's approach is similar to Walker's except *Bronn* develops an initial flaw distribution for the structure by regressing crack sizes found from a teardown inspection to zero time using crack growth analysis. *Kozin and Bogdanoff* present a statistical model for general application which characterizes the structural damage (fatigue, crack sizes, or wear) as a finite state-discrete Markov process. The models of Harris and Lim, Walker, Bronn, and Kozin and Bogdanoff all consider the effect inspections have on the structural reliability or on the probability of structural failure. Finally, *Albrecht* presents a method of calculating probability of fatigue failure by relating the fatigue loading histories for bridge structures to the fatigue properties of structural details.

**Probabilistic Fracture Mechanics
and Statistical Aspects of Fatigue**

The following discussion summarizes the individual papers in more detail.

Jouris performed a probabilistic analysis to assess the effects of various conservative assumptions used in Section III, Appendix G of the ASME Code. Section III, Appendix G uses linear-elastic fracture mechanics (LEFM) to ensure against the nonductile failure of the various parts of a pressure vessel at both operating and test conditions. Margins of safety are introduced in the design of the vessel due to deterministic LEFM methods prescribed in the Appendix. Inputs to the analysis are classified as either constants or random variables. The variability is accounted for by the use of probability density functions. The resulting calculated probabilities, according to Jouris, are to be used in a comparative way rather than as an absolute. Failure rates from 10^{-11} to 10^{-23} are estimated, depending upon how realistic the inputs are chosen. Jouris concludes that Appendix G vessels are indeed very conservatively designed.

Harris and Lim use probabilistic fracture mechanics (PFM) to estimate the probability of failure of large diameter nuclear vessel piping due to fatigue crack growth of initial crack-like defects. The PFM techniques were integrated in a computer code called PRAISE, which utilizes statified Monte Carlo techniques. Randomly distributed material properties were used along with two different initial crack depth distributions and three in-service inspection schedules. The main attention of the paper was directed to the investigation of the influence of in-service inspection on reliability. Conclusions were that the relative benefits of inservice inspection do not depend upon an exact knowledge of initial crack distribution. In addition, Harris and Lim show that leaks in large pipes are not very probable, but are much more likely to occur than a sudden double-ended pipe break.

Johnston has evaluated the material toughness and fatigue crack growth rates used to assess structural reliability. A review of statistically defined hyperbolic tangent reference toughness curves indicated that a tighter control on carbon, copper, and sulfur chemistry supported the hypothesis of improved brittle/ductile transition temperature for reference toughness. The investigation showed that K_{Ic} was able to be predicted from small-scale Charpy data. Assuming that C and m of the Paris power law fatigue crack growth rate expression are empirically related, Johnson was able to statistically show that C is lognormally distributed. Fatigue crack growth rates were evaluated in terms of stress ratio, temperature, and neutron irradiation effects. Lastly, an illustrative example was presented which showed the change in failure probability with number of accumulated cycles. Classical statistical methods were used throughout the paper as most analyses were based upon well-defined normal or lognormal distributions.

Trantina and Johnson combined the fracture linear-elastic fracture mechanics (LEFM) and fatigue S-N approaches to estimate the initial size of a defect which can lead to structural failure. An expression of S-N curves which incorporates the initial defect sizes is produced by integrating the cyclic crack growth rate. It is assumed that the scatter in fatigue life is dependent only on the scatter of the size of the strength-controlling defect. The assumption is based upon the Weibull weakest link concept that states the larger the amount of material sampled, the greater the probability that the material will contain a large defect. Three types of material data can be used in the analysis—cyclic crack growth data, fatigue S-N data, and initial defect sizes. Any two of these can be used to predict the third. The approach used is valid for cast metals and high-strength alloys only. As an example, a distribution of initial critical defect sizes are predicted for a nickel-based superalloy. The authors conclude that critical defect size distributions in components can be characterized so that material processing and inspection criteria can be based on required component lives.

Berens and Hovey have postulated a probability of crack detection (POD) as a function of crack length in order to be able to characterize the uncertainties of nondestructive evaluation (NDE). The true POD as a function of crack length will never be known exactly since NDE capabilities can only be demonstrated through experiment. However, statistical methods can be used to provide confidence limits on the true probabilities. Recent Air Force data are used to estimate the parameters of the statistical NDE model. NDE reliability experiments are statistically simulated and compared to theory. The authors conclude that from the large degree of scatter in POD estimates, such estimates are unreliable. Furthermore, both the magnitude and scatter in POD estimates are significantly influenced by crack size. A very large sample size would be required to reduce the "error" in the POD estimates in order to yield more precise crack length predictions of detection. Levels of 90 and 95 percent confidence are most difficult to obtain due to human factors and inspection hardware capabilities.

Ostergaard and Hillberry used a set of 68 crack growth tests for 2024-T3 sheet, published previously by Virkler, to define the constant C, and the exponent m in a modified Paris crack growth equation. The parameters C and m were obtained for each data set using a finite integral optimization routine which yielded the minimum error between predicted data points and test data points. A linear relationship was obtained between log C and log m by regression analysis. The statistical distribution of m-values was found to fit a 2-parameter lognormal distribution better than five other functions evaluated. The scatter about the regression line was represented by a 3-parameter lognormal distribution of a parameter, F, represented by the ratio of the predicted value of C from the regression equation to the value of C obtained from the finite integral optimization routine. The original data set was recreated within 10 percent confidence bands, by using a Monte-Carlo type prediction scheme for selecting m and calculating F. The results apply to one material conducted at one stress range in one environment (laboratory air) over a limited range of the da/dN versus ΔK curve, but the methodology could be applied to other materials and test conditions.

Walker presents a probabilistic model for evaluating the interrelationship of variables affecting crack-growth-based inspection programs. The model variables include fatigue cracking, crack growth, detectability of cracks, and a residual strength safety criterion. The model assumes all cracks start from the same size but will grow at different rates related to their probability of occurrence. For illustrative purposes, the fatigue life and normalized cracking rates were assumed to follow a lognormal distribution with a log standard deviation equal to 0.13. The normalized detectable crack size was also assumed to follow a lognormal distribution. The time to perform the first inspection is based on the time for a critical crack to reach a specified probability of occurrence, which was assumed to be 10^{-4}. Repeat inspections are based on the time required for the crack size with a probability of occurrence of 10^{-4} to reach the critical crack size. Probability distribution plots are used to aid in the calculations. Conservatism is achieved in this model because the cracks that impact the safety criterion grow at the fastest rates. Cracks that grow at average rates and less have very little impact on the repeat inspection interval.

Kozin and Bogdanoff's paper is one which considers the combined effects of fatigue, crack growth, failure, inspectability, and repair or replacement in assessing structural reliability. The method divides the damage process into a discrete number of repetitive periods of operation called duty cycles. The accumulated damage is modeled as a Markov process, where the damage is only considered at the end of each duty cycle. The damage can be modeled as either fatigue accumulated damage, a crack of a specified size, or a given amount of wear. In a stationary model, the repetitive duty cycles have the same fatigue loading severity and damage is incremented in unit jumps. This would be applicable where the fatigue loading is repetitive over relatively short periods of time. No discussion is presented on how one would modify the probability transition matrix (ptm) to account for variation in fatigue loading severity. The mean and variance of two damage conditions (for example, crack lengths) are used to define the number of duty cycles of equal severity (time) and the probabilities associated with the ptm. At the end of each duty cycle, the probability of a given damage size can be determined. Inspection and replacement or repair can be performed at the end of specific duty cycles which modifies the ptm for subsequent duty cycles. Examples are presented to illustrate the application of this methodology in assessing the reliability and maintainability of a component of a structure. Additional references have been published by the authors on related subjects and on extension of this methodology.

For aircraft fatigue loading, *Larson and Shawver* develop a relation between stress and g-counts (magnitude of vertical acceleration) by regression analysis performed on data sets of g-counts and stress measurements. The relationship can be applied to other aircraft that only record g-counts. The fatigue analysis is performed for increments of g-levels assuming a normal distribution of stress with a variance equal to the standard error of estimate from the regression analysis. Fatigue damage is calculated using Miner's rule and S-N curves for different levels of probability. A probability distribution of fatigue damage is obtained rather than a single discrete number as is currently done. The analysis is based on the assumption that all the damage is accounted for by the g-counts, that the damage is independent of the minimum stress associated with the g-counts, and that the damage obtained using Miner's rule is accurate for this application. Miner's method does not account for the effect of loading sequence which can have an effect on fatigue life. Also, the maximum g-count on each flight should be associated with the minimum stress during that flight to account for the effect of the ground-air-ground cycles.

Bronn's paper considers the combined effects of fatigue loading severity, crack growth, fracture, and inspection/repair in assessing the probability of structural failure. The initial damage condition is assumed to be a probability distribution of crack sizes. He illustrates how this initial distribution is determined from the results of a tear-down inspection. The cracks found in the tear-down inspection were converted to equivalent quarter circular corner cracks at fastener holes by crack growth analysis. The crack growth analysis was regressed for the service loading history to obtain the distribution of equivalent initial quarter circular corner cracks at time zero. This equivalent flaw size distribution was represented by a three segment Weibull probability distribution. The probability of crack detection for two inspection methods was based on results obtained from an Air Force program. An illustrative example shows how the fatigue loading severity and the type and frequency of inspections affect the calculated value of the probability of structural failure. The example illustrates the methodology as applied to one airplane but could be extended to apply to a fleet of aircraft.

Albrecht presents a method for calculating the probability of fatigue failure (cracking) for a structural detail on highway bridges. In his analysis, the load spectra, in the form of histograms, is replaced by a lognormal distribution of equivalent stress ranges. This transformation is applicable to bridge stress histories but cannot be applied to structures which are subjected to more complex load histories. The fatigue properties of a structural detail are represented by a lognormal distribution about a straight line plotted in logarithmic coordinates of stress versus cycles to failure. The equivalent stress range distribution is transformed to a lognormal distribution of design load cycles using the slope of the S-N curve. The fatigue reliability analysis is then determined by consideration of the difference between two lognormal distributions. Application of the analysis to designs meeting AASHTO Specifications indicate that different safety indexes are required for redundant load path structures than for nonredundant load path structures to achieve equal levels of failure probabilities.

Some of the papers discussed show how to characterize statistically some of the variables affecting structural safety and reliability. The available statistical analysis methodology is generally adequate for this purpose. What is generally lacking is sufficient and adequate data to define the probability and confidence level of the variables for specific applications. This is particularly true for predicting structural safety where very low probabilities of structural failure are required. More data need to be collected and evaluated for use in structural safety and reliability analysis.

Several models were presented for considering the interrelationship of variables affecting structural reliability and safety. One of the basic differences in the approaches is how one characterizes the damage state of a structure. In the probabilistic fracture mechanics approach, the structure is considered to contain a worst flaw when considering structural safety and a distribution of initial flaws when considering reliability or maintainability. In the fatigue approach it is assumed that cracks of some definable size are obtained after some amount of accumulated service time based on fatigue analysis or test data. The worst flaw approach is applicable for structures which are not readily inspectable in service or the critical crack size is below the reliably detectable crack size. For structures made from castings or weldments, an initial defect distribution is probably appropriate. Whether a fatigue damage model or a distribution of initial defects is the best way to characterize the damage state of other structures is unresolved at this time.

The analysis methods presented are adequate for making comparative evaluations of the effect some variables have on structural safety and reliability. The accuracy of the results depends on the assumptions inherent in the mathematical model and extent of available data. A more detailed analysis would consider other modes of cracking which are causes of structural failure, such as, corrosion, stress corrosion, fretting, service induced damage, etc. Also the effects of environmental exposure on the growth of defects and cracks would have to be considered. To provide more confidence on the use of this methodology for the design of structures, analyses should be applied to existing structures for which a large amount of service experience data are available. In this way, one could determine the correlation between predicted results and actual service experience.

J. M. Bloom

Babcock & Wilcox, Research and Development Division, Alliance, Ohio 44601; symposium cochairman and coeditor.

J. C. Ekvall

Lockheed-California, Burbank, Calif. 91520; symposium cochairman and coeditor.

Index

B

Binomial distribution, 81, 86, 87
B-model, 132, 134-138, 145
Brittle/ductile transition, 45-47

C

Charpy energy, 45, 47, 49-51
Coefficient of variation, 88, 89
Conditional failure rate, 39
Conditional leak probability, 31
Conditional cumulative probability, 31, 32, 35, 39
Confidence, 52, 53, 80, 117
  Bands, boards, interval, 81, 84, 111, 150, 154
  Bound limit, 52, 80-82, 86-93
Crack depth distribution, 24, 31, 32, 36-39, 41
Crack growth data, 56, 57, 59-61, 100
Crack growth law, 30, 43, 60, 63, 69, 100
Crack growth rate, 69, 70, 72, 73, 98-101, 109, 118
Crack propagation, 53, 55-57, 59-61, 69
Crack size defect distribution, 22-25, 62, 73-76, 85, 87, 122, 167
Crack tip opening displacement (CTOD), 47, 49, 50, 52, 53, 62, 64
Cycles-to-failure, 70, 71, 73, 148, 153

D

Density function (*see* Probability density function)
Damage data, 117
Distribution (*see also* Bionomical, Crack depth, Crack size, Expotential, equivalent flow size, Fatigue life expended, Gamma, Initial crack (flow) size, Initial damage, Lognormal, Normal, Poisson, Student t, and Weibull
Aspect ratio, 28
Duty cycle, 132

E

Equivalent flaw size, 162, 166-169, 174
Equivalent stress range, 185-188
Error analysis, 102
Error term, 101, 105
Exceedance frequency, 163
Exceedance probability, 162, 174
Exponential distribution, 27

F

Failure probability (*see* Probability of failure)
Failure criterion, 23
Failure rate, 24-26, 32, 33, 39
  Conditional, 39
Fatigue crack propagation, (*see* Crack propagation)

Fatigue life, 71, 72
  Expended, 148, 149
  Expended distribution, 158, 159
  Index, 147
  Finite intergral optimization, 101, 104, 106, 107, 110
Flaw size exceedance probability, 169, 175
Fracture toughness, 8-12, 15, 44-47, 49-51, 53, 64, 70, 77

G

Gamma distribution, 17
Goodness of fit, 83, 84, 110, 111

H

Hazard function, 24, 25, 133

I

Important densities, 9, 16, 17
Importance sampling, 9, 15, 16
Initial crack (flaw) size distribution, 23, 26, 27, 119
Initial damage distribution, 134, 138-141
Initial defect distribution, 76

J

J-intergral, 46, 49

K

Kolmogorov-Smirnov test, 111, 113, 114

L

Log logistics (log odds) model, 84, 86
Log normal distribution, 27, 30, 49-52, 61-64, 87, 117, 119, 158, 188

M

Margin of safety, 2, 9
Markov chain, 132, 133
Markov process, 132
Mean, 17, 30, 47, 49, 51, 61, 62, 85, 88, 89, 135-137, 149, 150, 154, 187, 188
Median, 30, 52
Miner's rule, 148, 152, 155, 185
Monte Carlo, 9, 16, 26, 85, 111

N

Normal distribution, 9, 17, 30, 49, 50, 63, 85, 110, 151, 155
NOE reliability, 81, 86, 87, 92
Nondetection probability (*see* Probability of nondetection)

O

Optimized probability method, 82

P

Probability density function, 9, 16, 24, 82, 110, 152
Probability distribution (*see* Distribution)
Probability of failure, 8-14, 22, 24, 25, 62, 64, 74, 121, 141, 170, 172-182, 187
Probability (*see also* Conditional cumulative, Conditional leak, and Flaw size exceedance), 17
  Cumulative failure, 22, 24, 31
  Detection (POD), 80-93, 117, 120, 124, 125, 127, 128, 139, 171, 172
  Double ended pipe break (DEPB), 30, 31
  Nondetection, 22-24, 26, 29, 170
  Replacement of, 138, 141
  Stress distribution, 151, 153

Stress exceedance, 163, 174
Survival of, 175, 176
Probability mass function, 133
Probability transition matrix, 132
Probabilistic fracture mechanics, 8, 21–23, 26
Poisson's distribution, 22, 23, 36
Post-inspection distribution, 22, 23, 36, 38

R

Random error, 149, 150
Random variables, 8, 9, 15–17, 33
Rank order number, 167
Regression analysis, 83, 85–87, 90, 91, 104, 105, 149, 151, 175, 176, 187, 188
Regression coefficients, 187, 188
Reliability function, 133
Reliability, NDE (see NDE reliability)
Reliability structural (see Structural reliability)
Residual strength, 164
Resistance curve, 46, 188

S

Safety criterion, 117, 121–123, 125, 126
Safety factor, 2
Safety index, 187
Severity factor, 8–11

S-N curve, 72, 152, 153, 155–157, 159, 185, 187
Standard deviation, 30, 61, 62, 85, 89, 109, 187, 188
Standard error, 86, 90
Standard error of estimate, 150, 152, 154
Standard normal variable, 152, 187
Stress concentration factor, 74, 76
Stress intensity factor, 9, 10, 43, 69–71, 73, 77
  Threshold (see Threshold stress intensity factor)
Stress intensity range, 69–72, 75
Stress range histogram, 185–187
Structural reliability, 22, 186
Student-t distribution, 154

T

Time to failure, 134
Threshold stress intensity factor, 69, 100, 101, 103
Tolerance limit (board), 47, 49, 51–53

V

Variance, 16, 17, 51, 83, 135–137, 149–151, 154

W

Weibull distribution, 49, 50, 64, 73, 74, 77, 83, 169